Economic REVITALIZATION

TO OUR PARENTS

LaVerne Schroeder Hufnagel

IN MEMORY OF
Walter Frederick Hufnagel
Thomas Watkins Leigh
Henrianne Leigh Miles

Economic REVITALIZATION

Cases and Strategies for City and Suburb

Joan Fitzgerald
Nancey Green Leigh

SAGE Publications
International Educational and Professional Publisher
Thousand Oaks ▪ London ▪ New Delhi

For information:

Sage Publications, Inc.
2455 Teller Road
Thousand Oaks, California 91320
E-mail: order@sagepub.com

Sage Publications Ltd.
6 Bonhill Street
London EC2A 4PU
United Kingdom

Sage Publications India Pvt. Ltd.
M-32 Market
Greater Kailash I
New Delhi 110 048 India

Printed in the United States of America

Library of Congress Cataloging-in-Publication Data

Fitzgerald, Joan, Ph.D.
  Economic revitalization: Cases and strategies for city and suburb /
Joan Fitzgerald, Nancey Green Leigh.
    p. cm.
Includes bibliographical references and index.
  ISBN 0-7619-1655-5 (c) -- ISBN 0-7619-1656-3 (p)
  1. Urban economics. 2. United States--Economic conditions. I. Leigh,
Nancey Green. II. Title.
  HT321 .F57 2002
  338.973'009173'2--dc21

                                                    2001005796

            06   10   9   8   7   6   5   4   3

Acquiring Editor:	Margaret Seawell
Editorial Assistant:	MaryAnn Vail
Production Editor:	Diana E. Axelsen
Typesetter:	Siva Math Setters, Chennai, India
Indexer:	Janet Perlman
Cover Designer:	Michele Lee

Contents

Acknowledgments

The idea for this book was conceived on a Sunday bike ride we took one summer along the lakefront from central Chicago to the inner-ring suburb of Evanston. The actual writing of this book has taken us individually and jointly to many other cities and suburbs, and we would like to acknowledge the support of many along the way who have helped us realize its completion.

We would like to thank Catherine Rossbach, our first editor at Sage, who provided encouragement and support throughout the writing process. We also thank Sam White of the University of Wisconsin at Milwaukee and Dick Bingham of Cleveland State University for reading drafts of early chapters and providing helpful comments and encouragement. Our current editor at Sage, Marquita Flemming, has provided valuable help during the final stages of completing the manuscript. Wim Wiewel, former director of the Great Cities Institute at the University of Illinois at Chicago, kindly provided Nancey Green Leigh with a visiting research fellow position for her research project on the office and industrial property markets of Chicago, which led to the fateful bike ride.

As a faculty fellow at the Great Cities Institute and a faculty member of the College of Urban Planning and Public Affairs at UIC, Fitzgerald thanks Wiewel and Great Cities Associate Director Lauri Alpern for their support. Kimberly Gester, Tom Stribling, and Jean Templeton provided research assistance for parts of several chapters. The late Rob Mier and Bennett Harrison were sources of personal and professional inspiration on placing social justice at the forefront of economic development.

Fitzgerald thanks staff and board members of the Jane Addams Resource Corporation for providing access to records, staff, students, and interviews for the case study in Chapter 2. Board Chair Hal Baron, Executive Director Michael Buccitelli, and Associate Director Anita Jenke Flores commented extensively on the manuscript. Jon Soderstrom, head of the Office of Cooperative Research, was particularly helpful in providing background information on Yale's involvement in biotechnology and in making contacts with other key sources. Debra Pasquale, president of Connecticut United for Research Excellence, Inc. (CURE), also provided lengthy interviews and introductions to sources in state and city government.

Leigh wishes to acknowledge the Lincoln Institute on Land Policy for supporting her research project "Promoting More Equitable Brownfields," which is drawn on in parts of Chapter 3, "The Brownfield Redevelopment Challenge." She also is grateful for the brownfield research assistance of Georgia Tech master's student Crystal Welsh, and doctoral student Sarah L. Coffin.

Fitzgerald is appreciative of comments and insights on Chapter 4, "Industrial Retention: Multiple Strategies for Keeping Manufacturing Strong," from Jon DeVries, Donna Ducharme, Bob Giloth, Jim Lemonides, Kari Moe, Dennis Vicchiarelli, and Luke Weisburg, all of whom were involved in economic development and industrial retention in Chicago at various times. Ben Wolters provided input and contacts on Seattle's industrial retention efforts.

For Chapter 5, "Commercial Revitalization in Central Cities and Older Suburbs," Leigh is grateful for the research assistance of Georgia Tech student Jennifer Ball. She also thanks the staff of Sandy Springs Revitalization, Inc., John Cheeks, Alan Steinbeck, and Donna Gathers, for their assistance in preparing the chapter's case study through interviews, providing materials, and comments on the draft text.

For assistance in research that is drawn on in Chapter 6, "The Reuse of Office and Industrial Property in City and Suburb," Leigh is thankful to Georgia Tech master's students Jonathan Hoffman and Peter Vaughn. She also gratefully acknowledges the support of the Lincoln Institute on Land Policy for her research project on the reuse of office and industrial vacant land in central cities. In particular, Leigh wishes to thank Rosalind Greenstein, director of Lincoln's Program on Vacant Land, for the interest and encouragement she has shown in her work.

Chapter 7 relied heavily on long interviews with several staff members of the Seattle Jobs Initiative and in several other Seattle organizations. In particular, Fitzgerald thanks Mary Jean Ryan, director of Seattle's Office of Economic Development, for candidly sharing progress, successes, and frustrations of the evolving Seattle Jobs Initiative. Several people offered valuable comments on the chapter: Bob Giloth, program officer at the

Annie E. Casey Foundation; Alex Schwartz, associate professor at New School University; and Bob Watrus, policy analyst at the Northwest Policy Center at the University of Washington.

Finally, we acknowledge the support of our respective institutions, the City and Regional Planning Program at the Georgia Institute of Technology, the University of Illinois at Chicago, and Northeastern University. Jeann Greenway, at Georgia Tech, provided the last critical assistance that was needed to get this manuscript out the door.

Introduction:
Why We Wrote This Book

A New Perspective on Local Economic
Development Planning

Most of our students, whether recent college graduates or seasoned practitioners, view planning as a way of changing the world. Many students and practitioners tell us of witnessing or being victims of job loss, gentrification, neighborhood deterioration, or environmental racism as what motivated them to want to practice economic development planning. Yet they also express disappointment at how little impact they have in achieving these "idealistic" goals of planning once they become practitioners. Indeed, given the international forces affecting our economies, it is easy to lose faith in economic development planning as being capable of accomplishing much more than "tinkering at the margins." We can encourage the reuse of an abandoned property in an inner-ring suburb but cannot quell the larger tendency to develop greenfield sites. We can create a few effective job-training programs, but we cannot improve the increasing polarization of wages. We can keep a few manufacturers from leaving the city, but we cannot stop the exodus of manufacturing to the suburbs or offshore locations.

Hearing these comments for many years, we began to wonder—are we giving the future generation of economic development practitioners unrealistic expectations of what planning can do? After many discussions, we concluded that their expectations are not unrealistic, but that we often fail to integrate into our teaching a framework for incorporating values such as social justice into the most common strategies students will be implementing as practitioners. We teach theory. We teach methods. We teach strategies. However, we do not teach how to incorporate the key values that motivate our students in the first place into the most common economic development strategies.

Furthermore, we often teach economic development as if it will only be practiced in inner cities, when in fact an increasing number of practitioners will work in suburbs or edge cities, or in all three environments, over the course of their careers. Because one-half of the U.S. population now lives in suburbs, there is a need for an economic development book that specifically discusses practice in suburban as well as inner-city settings. Economic activity originally followed population shifts to the suburbs in order to provide commercial services. However, in the last 15 years in particular, the continued movement of other types of businesses has transformed many suburban bedroom communities into independent economies. Moreover, as population shifts continue their path of decentralization, further suburbanization and exurbanization has resulted in first-ring suburbs experiencing problems of decline requiring economic development assistance much like those of inner cities. Often the same strategies and methodologies are used in both cities and suburbs, although actual program implementation is likely to be different.

Although *Economic Revitalization: Cases and Strategies for City and Suburb* is aimed at students and practitioners of economic development planning who seek to foster stronger economies and greater opportunity in inner cities and older and inner-ring suburbs, it is also meant to assist planners in thriving new towns and suburban communities seeking to avoid future economic decline as their communities mature. We hope to provide a useful reference for students, planners, and other practitioners that identifies components of essential economic development strategies, case studies of how these strategies are being used in inner-city and suburban settings, and issues that emerge in their implementation.

In several of our case studies, we provide an in-depth focus of implementation so readers can gain a better understanding of how the planning process works. Specifically, the cases of implementation reveal the political nature of the planning process and the types of trade-offs often made. Sometimes these trade-offs involve compromising on issues of equity and sustainability. However, sometimes windows of opportunity for putting one or the other at the forefront can be negotiated. Another reason for our focus on implementation is the insights it reveals to planners seeking to adopt "best practice" programs from other cities and states. Seldom is best practice simply a set of program features identified at a given point in time. Best practice programs usually are the result of frequent adaptation over months or years of implementation. We believe planners need a better understanding of the process of implementation to adapt programs to fit the needs and political realities of specific places.

We begin this book, in Chapter 1, by taking stock of what has gone on in the name of economic development planning and practice over the past 6 to 7 decades. We then offer a (re)definition of economic development planning to guide future practitioners that emphasizes equity and sustainability as goals. We define local economic development: *preserving and raising the community's standard of living through a process of human and physical infrastructure development based on the principles of equity and sustainability.*

The imperative for our emphasis on equity and sustainability cannot be missed by anyone scanning today's spatial economy. Although many of the nation's central cities are experiencing a renaissance, poverty still persists. Furthermore, with the majority of the nation's poor living in mixed-income city and suburban neighborhoods, urban poverty can no longer be viewed as an inner-city ghetto problem (Blank, 1997). As we elaborate in the next section, major and minor metropolitan areas throughout the nation suffer from all the attendant problems of urban sprawl: leap-frog development, pockets of decay, jobs/housing imbalances, deteriorating air and water quality, and clogged freeways and arterials. The largest metro areas all have bedroom suburbs that have become edge cities overwhelmed by business growth and commuting nightmares.

Central Cities and Older Suburbs:
Examining the Common Ground

The picture of the United States that emerged from the 1990 Census gave us a new image of ourselves: We had become a suburbanized nation. The majority of Americans lived not in cities or in rural areas but in suburbs. At the beginning of the 1990s, roughly one-half of the population lived in suburbs while the remaining one-half was split fairly evenly between rural and central-city areas. Much was made of this new, seemingly unifying and homogenizing trend. As a nation, we have begun to realize, though not as quickly as we should have, that while suburban form dominates the American landscape and economy, the image of the prosperous, family-friendly, and secure suburb does not uniformly apply. All suburbs do not equally represent and share in the so-called American (read, Suburban) Dream. In particular, older suburbs often find themselves sharing the problems of the inner city. Indeed, at the forefront of challenges in this century for economic development practitioners, and planning in general, are the persistent problems of inner cities to which the parallel problems of increasingly minority and poor inner-ring suburbs must now be added.

A recent study of places left behind published by the U.S. Department of Housing and Urban Development (1999a) confirms that many older suburbs are increasingly poor, with rates of job loss as high, and in some cases higher, than inner-city neighborhoods. Pockets of inner-ring suburban poverty are documented in Minneapolis, San Francisco, Rochester, Philadelphia, Boston, Cleveland, and other cities. The patterns of out-migration, disinvestment, and tax base erosion parallel that which inner cities experienced in previous decades.

Former Minnesota State Representative Myron Orfield is a leading advocate of metropolitan-level planning solutions for addressing inequality. Orfield (1998) writes that 20% to 30% of the nation's metropolitan area populations "live in low-tax-base inner suburbs and satellite cities that are rapidly looking more like central cities" (p. 1). Furthermore, when this proportion of the population is added to that of the nation's population in central cities, "the total number of residents disadvantaged by regional polarization comes to well over 65%" (Orfield, 1998, p. 1). The remaining percentage who reside in the stable, secure suburbs have high tax bases and few social spending needs. However, these suburbs dominate a metropolitan region's growth and consume a disproportionate share of the region's highway, sewer, and other infrastructure spending.

Orfield cites two indicators—property tax base and job growth—to show the diversity and polarization of the suburbs and uses Los Angeles as an example. The average property tax base per household across the region was $168,382 (in 1996), compared with $147,307 for the City of Los Angeles. Eighty-six suburban cities had average property tax bases per household lower than the City's, and 37 of these were 70% or less (including places such as Compton, Bellflower, San Bernardino, and Fontana). Thus, low property tax bases are associated with suburban communities that have the greatest social needs. Eighteen communities had tax bases per household exceeding $400,000. During the 10-year period between 1986 and 1996, the metrowide average property tax base per household increased 9.3%. The City of Los Angeles' increase was less, at 7.1%, but still positive. However, 17 communities in the Los Angeles region had property tax base declines greater than 15%.

Job growth patterns across the Los Angeles region also reveal similar intrametropolitan disparities: Between 1990 and 1994, the region as a whole experienced a 12.3% decrease in jobs per 100 persons, such that there were 42.4 jobs per 100 persons in 1994. The greatest job losses were associated with the inner-suburban and satellite cities (for example, Lynnwood dropped from 25.2 to 17.6 jobs per 100 persons, and Highland from 17.9 to 10.5). The cities with the highest jobs per capita in 1990 experienced the greatest percent increase in jobs (for example, El Segundo saw an increase from 327.6 to 351.7 jobs per 100 persons, while Irvine experienced an increase from 108.1 to 121.4 jobs per 100 persons).

Orfield (1998) observes the following:

> There is an illusion that poverty, social instability and community decline stop at central-city borders and that U.S. suburbs are monoliths of affluence, social harmony and political unity. In truth, the suburbs are as diverse—and as plagued with problems—as the central cities they surround. (p. 1)

To differentiate between suburbs, it is necessary to define the older (also called inner-ring or first-ring) suburbs, many of which developed in the postwar period and are now experiencing economic distress. The University of Minnesota's Design Center for American Urban Landscape's (UM-DCAUL) definition of inner-ring suburbs is the most widely cited (Design Center for American Urban Landscape, 1999). Inner-ring suburbs are those that developed between 1945 and 1965; they are a group of bedroom communities and small towns sprouting up after World War II "that serviced and relied on the economic prosperity and cultural amenities of the central cities."[1]

UM-DCAUL has identified three characteristics of inner-ring suburbs. First, they are close neighbors to the central city. Second, they were once the edge of the metropolitan area and continue to exhibit some edge qualities. Third, they are communities with their own distinctive characteristics. The inner-ring suburbs are now the middle of the metropolitan area, or the "new urban pivot point."

The extent of poverty and other indicators of economic distress, as well as the changing demographics of the inner-ring suburbs, are not captured very well in existing and easily accessible databases. The boundaries of many suburbs are not separate political jurisdictions— either city or county—for which data are typically collected. Therefore, their circumstances are not neatly reflected in readily available city and county data sources such as the diennial Census of Population or the annual County Business Patterns.

On the other hand, the persistence of poverty in central cities has a long history of documentation, although the record of planning and policy solutions addressing it has been generally unsuccessful. While the current economic boom has meant good news to many U.S. cities, over-all economic growth does not guarantee poverty reduction because of falling wages for unskilled jobs (Blank, 1997). A recent report of the U.S. Department of Housing and Urban Development (1999b) documents economic and social gains for many cities in several areas, including job growth, home ownership, population, fiscal strength, and declining crime rates. Unfortunately, the report continues to describe a darker side to these trends:

> Despite the positive news, too many cities and pockets of concentrated poverty are being left behind in urban America's impressive comeback. These cities continue to suffer from the challenges of population decline, loss of middle-class families, slow job growth, income inequality, and poverty. (U.S. Department of Housing and Urban Development [HUD], 1999b, n. p.)

Although cities overall are faring well, serious population declines continue to plague about 20% of our nation's 539 central cities. High unemployment (50% or more above the national rate) affects 17% of central cities. Nearly 1 in 3 central cities had poverty rates of 20% or more in 1995, and even in cities with lower rates, poverty remains concentrated in selected urban neighborhoods. Poverty rates of 20% or more can be found in 170 central cities in 34 states and the District of Columbia.

Central Cities and Older Suburbs: Common Planning Practice?

We think there are local economic development strategies that can address similar problems of large inner cities, small and mid-sized cities, and inner-ring suburbs. Several economic development texts discuss how to analyze local economies and how to choose locally appropriate strategies. However, the truth is, the same strategies are lauded indiscriminately everywhere. The current roster includes industrial retention, business attraction, tourism, commercial revitalization, small business development, and casino and sports stadium development. As we note in Chapter 1, the manner in which these strategies are adopted does not reflect a carefully thought-out local economic development strategy, but rather a following of the latest fad.

Our Goals in Writing This Book

Based on our interaction with students and practitioners of economic development over many years, we believe there is a need for an economic development text rooted in the planning values of social justice and sustainability. We hold that good local economic development practice should specifically take into account the distributional consequences of resulting growth and change.

Guided by this perspective, we focus on six common economic development strategies that most local economic development planners and practitioners will be charged with implementing over the next decades: industrial retention, brownfield redevelopment, sectoral (or targeted industry) strategies, commercial revitalization, industrial and office property reuse, and workforce development.

We have three goals in writing this book. Our first goal is to examine how the principles of economic and social equity, as well as environmental sustainability, can be built into common economic development strategies. To adhere to these principles, we adopt an alternative definition of economic development (elaborated on in Chapter 1) to that of the conventional business interest-driven and wealth creation approach currently dominating the field.

Our second goal is to discuss how common urban economic development strategies can be applied in older suburbs in addition to cities. Although this goal is aimed at helping the field of local economic development become more effective in solving the problems of inner cities and older suburbs, we believe our discussion of the six strategies can also help practitioners in thriving new towns and suburban communities prevent economic decline as their communities mature.

Our third and final goal is to inform the teaching and practice of economic development by focusing on implementation and the politics of local economic development. We suggest that it is not the strategy per se, but rather how it is implemented that determines its outcomes. The cases of implementation in these chapters reveal the political nature of the planning process and the types of trade-offs on issues of equity and sustainability that often must be made. We also focus on implementation because of the insights it reveals to planners seeking to adopt "best practice" programs from other localities. We hope that this book will help them to do so with a practical and theoretically grounded presentation of the whys, whens, and hows of employing such strategies to create growth while striving for equitable distribution and long-term sustainability.

Note

1. Elsewhere, the time frame for the formation of the inner-ring suburbs has been extended to 1977 (Muschamp, 1997). The postwar frame of reference for the formation period of these suburbs, however, is most applicable to newer (and/or sunbelt) metropolitan regions. Throughout the country and, in particular, the Midwest and Northeast regions, inner-ring suburbs frequently date back to the turn of the century and the advent of streetcar suburbs.

1

Redefining the Field
of Local Economic Development

As a profession and an academic field within urban planning, economic development is relatively young. We begin this chapter with a brief history of economic development practice and theory. Then, we identify key economic development issues in inner cities and inner-ring suburbs and discuss tools and practices aimed specifically at urban or inner-city revitalization. We conclude that there is a need to reconsider how economic development is defined. Without such a reconsideration, the practice of economic development will continue to be as hit-or-miss and inequitable as its critics assert. We make a case for a reorientation of economic development practice for central-city and inner-ring suburbs that works toward achieving a higher degree of social equity and sustainability.

A Brief History of Local Economic Development Practice

Historically, we can identify five broad trends or phases of economic development practice at the state and local levels.[1] These phases are both chronological and overlapping. In Phase 1, beginning in the 1930s, practitioners mainly sought to attract industry with relatively crude measures intended to reduce production costs via measures such as tax abatements, land assembly and write-down, and public

9

infrastructure. Economic development was not yet a subject of academic inquiry. Beginning in the 1960s, in what we are calling Phase 2, serious political critiques of economic development practice emerged from both activist and scholarly critics. Advocacy planning emerged. In Phase 3, state and city economic development efforts added programs to stimulate export markets for locally produced goods, as well as labor market and research and development (R&D) enticements. As the practice became more sophisticated, so did political critiques of planning practice. Consequently, equity planning emerged. In Phase 4, starting in the 1980s, environmental sustainability as well as equity became a concern of critics and some practitioners. In Phase 5, the present, there is renewed concern about sprawl and interest in metropolitan or regional strategies, as well as an intensification of market-driven solutions. We suggest, however, that aspects of earlier trends still survive and are manifested in later phases.

Phase 1: State Industrial Recruitment

Phase 1 began with the efforts of individual states to attract industry in the 1930s. The evolving economic development practice attempted to create a good business climate through tax abatements, loan packaging, infrastructure, and land development, as well as other efforts to reduce the cost of production for firms. In essence, this marked the beginning of corporate welfare, through which public funds were directed to private sector firms to influence their location decisions. This approach reached its zenith in the late 1960s with a proliferation of predominantly suburban and rural industrial parks serving as locations for manufacturing firms.

The private sector practitioners in this period typically were real estate developers, chambers of commerce, and utility company location specialists.[2] In the public sector, they were mainly state government officials in nascent development agencies and extension agents from local universities. These practitioners sought legitimacy by joining professional associations such as the American Economic Development Council. Scholars in the disciplines of urban and regional planning, sociology, geography, and political science began documenting and analyzing the practice.

Two distinct perspectives emerged to explain and inform this early development practice: a regional and community development focus drawing on international development theories and experiences and an industrial location focus extrapolated from theories of firm behavior.

What characterizes these two perspectives is their concentration on identifying the causes of regional growth and development and, to a lesser degree, the extent to which local efforts can alter development and its orientation. Within these perspectives, a community's economic development potential was seen as merely a function of what it imports and exports.

The overarching principle underlying this phase of practice was that "greasing the skids" for business by offering tax abatements and other incentives is the appropriate focus of economic development. Marris (1982) suggests this view represents a corporatist paradigm in which "the primary task of government is to make the most of the land, labor, skills, raw materials, infrastructure and social amenities within its jurisdiction" (p. 30). Government's role is to act in an entrepreneurial capacity to ensure the profitability of firms within its jurisdictions. In this view, social goals incompatible with competitive success are unrealistic and self-defeating and therefore do not warrant consideration (Marris, 1982). This perspective defines the "traditional" economic development activity still being practiced today.

Phase 2: Political Critiques of Local Economic Development Activity

The second theoretical perspective on local economic development emerged in the late 1960s, following a series of events that raised questions about the impacts of planning practice. Although industrial development activity still dominated practice, the combination of rapid urban decline with a wholesale exodus of manufacturing from urban areas, large-scale civil disturbances, and the failure of urban redevelopment efforts forced on the table a new set of questions about what economic development practice should try to achieve. The focus of economic development analysis shifted from examining how to implement various techniques and strategies to questions of who was paying for, and who was benefiting from, the practice. The actors in economic development were no longer seen as faceless firms or neutral "rational" planners but as political agents.

These analyses of the political economy of local development focused on who participates in the process and on their motives for doing so. Researchers observed local businesspeople in proactive roles, often in collusion with local government officials (see Walton, 1979). In an influential article, "The City as a Growth Machine," Harvey Molotch (1976) suggests that economic development activity was led by local landholding elites interested in increasing the value of their property. Because local elites have fixed capital investments (real estate interests, utilities, capital-intensive manufacturing facilities), they participate in the development process to protect these investments.

Economic development planning, then, was justified in the name of job creation but practiced in the interests of wealth creation for these elites. Questioning one of the main justifications of local economic development practice—job creation—Molotch (1976) argued that the activities pervasive among practitioners did not create jobs but merely transferred them from one location to another. Other critics argued that tax abatements and other inducements offered to industry under the guise of economic development paid firms for what they would have done anyway. They pointed out that economic pressures emerging in the late 1970s—international competition and a sagging world economy—spurred an industrial restructuring that created a new spatial division of labor (Massey, 1984; Walker, 1984). Business attraction, the main activity of economic development, became referred to derisively as "smokestack chasing." Evidence began to accumulate that the promised jobs often were not created and the tax abatements and other infrastructure improvements provided by cities and states nullified any tax revenue increases that might have been realized (see Harrison & Kanter, 1978; Goodman, 1979; Jacobs, 1979; Vaugh, 1979).

This perspective revealed a fundamental contradiction in local economic development practice between the interests of cities and the interests of businesses within them. While cities have as their primary objective *stability* of employment and the tax base, firms seek *mobility* to produce in the lowest cost or highest profit location. When these interests coincide, such as with the locally dependent elites defined previously, public-private partnerships are formed to protect the mutual interests of places and business. When they do not coincide, cities may take action to prevent firm mobility, but such cases are rare. The more likely scenario is for economic development practitioners to act in the interests of the business sector, often to the point that these actions become impossible to justify given their limited impact on job creation or increasing the tax base.

The use of public resources for promoting economic development has largely gone unaccounted for, but the relatively few studies that have attempted to evaluate the effectiveness of economic development activities (usually the more easily measured fiscal incentives) have not demonstrated strong results. Academics in the economic development field have been pointing this out for some time, with little impact on the practice. The issue was raised for a much broader audience in a late 1990s *Time* magazine cover series (Bartlett & Steele, 1998) on corporate welfare, defined as the indiscriminate use of public resources for private sector gain. One example presented was the 1997 incentives given by the City of Philadelphia and the state of Pennsylvania to a Norwegian company (Kvaerner ASA) to reopen a portion of the closed Philadelphia Naval Shipyard. The article highlights the fact that the investment will take more than 48 years to be earned back by the taxes its workers will subsequently pay. The company received $307 million in incentives, or $323,000 for each of the 950 jobs it is creating. Workers in these jobs will be paid around $50,000 a year. In the *Time* series, Bartlett and Steele (1998) concluded the following:

> Two years after Congress reduced welfare for individuals and families, this other kind of welfare continues to expand, penetrating every corner of the American economy. It has turned politicians into bribery specialists, and smart business people into con artists. And most surprising of all, it has rarely created any new jobs. (p. 38)

In summing up this incentive-based economic development approach, Krumholz (1991) states: "In effect, the public economic development practitioner becomes an arm of the private developer, the success of the latter is a measure of the effectiveness of the former" (p. 293). Further, we add that the success has been measured in only the shortest of terms. What benefits the *current* private developer is all that is calculated. Whether, for example, the land use planning and design of a facility can be used easily by another business should the current business fail or move is seldom considered. Indeed, a major characteristic of postwar economic restructuring has been the shift to shorter and shorter business life cycles, as well as greater firm mobility. Public incentives and other efforts to facilitate private development go toward increasingly transient "successes."

Changing the practice is not easy. Good Jobs First, a project of the Institute on Taxation and Economic Policy (ITEP),[3] serves as a watchdog on corporate welfare. They estimate that since 1987, New York City has spent over $2 billion on retention or attraction incentives provided to 66 corporations. In 1994, Greg LeRoy, the founder of Good Jobs First, documented the proliferation of state incentives in *No More Candy Store* and described efforts of several states to develop alternatives, or at least to build in mechanisms for compensation (referred to as clawbacks) for companies that do not honor commitments made as part of these deals. An update to the report, *Minding the Candy Store* (Hinkley, Hsu, LeRoy, & Tallman, 2000), documents the continued proliferation of subsidies and the lack of accountability in documenting that they meet employment goals.

Although several states have entered into noncompetition agreements, these are easily broken when a firm dangles an offer of jobs in front of state officials (Burstein & Rolnick, 1994). For example, Governor James Florio of New Jersey broke a 1991 New York, New Jersey, and Connecticut "nonaggression" pact within months of signing (Mahtesian, 1994). This state of affairs is like a prisoner's dilemma, in that a city or state that stops using incentives is at a distinct disadvantage to those who continue (Bucholz, 1998). While incentive-based economic development practice still dominates, an alternative practice emerged, which is chronicled in the next section.

Phase 3: Entrepreneurial and Equity Strategies

As the promise of the traditional economic development strategies faded among practitioners, two separate movements emerged, marking the shift into the third phase of theory and practice of local economic development. The first was a shift from the supply-side industrial attraction focus to a more entrepreneurial focus, which occurred in both state and local economic development practice (Eisinger, 1988, 1995). These "second wave" strategies shifted the emphasis of economic development to promoting the development of new businesses and industries, particularly in high-tech sectors. The approach was still definable within the corporatist approach, with the public practitioner still viewed primarily as the agent of the private sector. The second movement advocated a set of alternative place-based strategies that focused on issues of equity and redistribution. This movement would later become known as equity planning.

Following the success of Silicon Valley, many states began programs in which university-based research was to be used to promote high-tech development (see Luger & Goldstein, 1993; Osborne & Gaebler, 1993). This approach is best summarized by Saxenian (1985):

> In 1983, thirty-three U.S. states spent over $250 million on R&D facilities to speed the growth of high-technology industries within their borders, while thousands of local governments schemed to devise the optimal package of tax breaks, incentives, and regulatory relief in order to lure entrepreneurs and innovative firms to their turf. Attracting high tech is *the* economic development strategy of the 1980s. (pp. 125–126)

States and cities also began initiatives such as international trade promotion, venture capital funds, small business development, and other programs to promote entrepreneurialism (see Eisinger, 1995).

Arising within this more entrepreneurial approach were efforts to be more discriminating about when to offer subsidies. The City of Chicago pioneered in efforts to analyze when to offer subsidies by using a cost-benefit methodology developed at the UIC Center for Urban Economic Development (see Wiewel, Persky, & Felsenstein, 1994). Further, some cities became more aggressive in responding to companies not honoring their hiring or other commitments, demanding "clawbacks" and even filing lawsuits (Mier & Giloth, 1993; LeRoy, 1994). In 1984, the City of Chicago filed a suit against Hasbro-Bradley for breach of contract when the company announced it was closing its Playskool plant that employed 1,200 workers. The company had expanded its Chicago facility in the late 1970s with $7 million in state low-interest loans and additional city funds. A settlement was reached in 1985 in which 100 workers were retained to operate a job-placement program. Inducements of $500 were offered to employers who hired the workers. Further, the company agreed to a $2.3 million severance package negotiated by union and nonunion workers. Such efforts were, and are, the exception rather than the rule.

Although this new wave of economic development strategies allows states and cities to be more proactive and strategic in shaping the type of development that takes place, their dominant focus is still promoting economic *growth*. Beginning in the late 1970s and early 1980s, a growing number of practitioners and academics in policy and planning began urging the profession to confront issues of socioeconomic inequality and to become advocates for those left out of the development process—in both participation and outcome. A motivating force behind the emergence of this equity planning perspective was the realization that the economic and political coalitions that drove urban economic development planning in most cities had changed.

In the 1980s, progressive regimes in several cities promoted strategies in economic development, housing, and community development to better serve the interests of low-income residents (see Clavel, 1986). In the early 1990s, this approach became known as equity planning, defined by Metzger (1996) as

> A framework in which urban planners working within government use their research, analytical, and organizing skills to influence opinion, mobilize underrepresented constituencies, and advance and perhaps implement policies and programs that redistribute public and private resources to the poor and working class. (p. 113)

This approach introduces redistribution of wealth into the goals of planning. Academic planners who identified with the equity perspective conducted studies of urban development focusing on the extent to which development policy was worked out through broad participation or as part of a closed-door process conducted by special commissions or mayoral-appointed study groups (Fitzgerald & Cox, 1990). In *The Contested City*, John Mollenkopf (1983) concluded that the old growth coalition composed of businesses and economic development planners was dead and advocated for its replacement by community-based coalitions. As Krumholz and Clavel (1994) noted, "the mainstream planning that serviced city development was no longer adequate to significant segments of city populations; namely, the poor, minorities, and elderly people with few resources who remained trapped in central cities as wealthier people moved out" (p. 4).

Economic development under the administration of Chicago Mayor Harold Washington (1983–1987) provided the first comprehensive example of how racial and social justice could be the underlying goal of a city's approach to economic development (Mier, Moe, & Sherr, 1986). Although some of the methodologies and tools equity planners used were similar to traditional planning, what distinguished this emerging practice was its framing of development questions. For example, Robert Mier (1994), Chicago's Commissioner for Economic Development under Mayor Harold Washington, defined the dominant themes of economic development planning during the Washington years as "targeting the work needy, job development (over real estate development), neighborhood development (over downtown development), business retention and expansion (over business attraction), small business development and targeting of such resources as purchasing" (p. xvii).

That is, equity planning asks that planners consider a different set of questions in evaluating economic development strategies—particularly the question of who benefits and who pays (see Krumholz & Forester, 1990). It also expands the notion of who participates in local economic development decision making beyond the public-private partnership, to include neighborhood organizations, civic groups, and labor unions. Community-based organizations were important partners with city government in developing and implementing economic development strategies in Chicago and Cleveland (see Clavel & Wiewel, 1991; Mier, 1993, 1994; Krumholz & Clavel, 1994). Indeed, community-based organizations have become important actors in economic development in many cities (see Clavel, 1986; Vidal, 1992; Simmons, 1994; Harrison & Glasmeier, 1997). Labor unions are also becoming engaged in local economic development issues and influencing electoral politics, particularly in Los Angeles and several other California cities (see Myerson, 2001).

In addition to expanding participation in the planning process, equity planning introduces new ways of examining old problems such as identifying the race and gender implications of economic development strategies and programs. Equity planners, for the most part, are still a minority voice in economic development practice. To make any inroads on dispersing or redistributing the benefits of development, they have to work out compromises with city governments more committed to downtown development and attracting middle-class residents to targeted neighborhoods.

Phase 4: Sustainability With Justice

In addition to adding equity as a goal of economic development planning, there have been increasing calls for more environmentally sensitive economic development over the last two decades. In an article calling for green, growing, and just cities, Scott Campbell (1996), a planning professor at the University of Michigan, argues the planner must reconcile three conflicting interests: "to 'grow' the economy, distribute this growth fairly, and in the process not degrade the ecosystem" (p. 297). Practice in the real world restricts planners to the point that they "usually represent one particular goal—planning perhaps for increased property tax revenues, or more open space preservation, or better housing for the poor—while neglecting the other two" (Campbell, 1996, p. 297).

Campbell depicts a triangle with the three points of "economic growth" and "environment" at the base and "justice" at the top (see Figure 1.1).

The triangle's points represent divergent interests leading to three fundamental development conflicts over property, resources, and development. With two of the conflicts—property and resources—the basic contradiction is the private sector resisting regulation of both property and natural resource use (Campbell, 1996). It does so while simultaneously relying on regulation to keep the economy flowing and conserve resources for meeting present and future demand. Balancing this contradiction, we note, is the essence of traditional local economic development's focus on the public-private partnership.

The third conflict, "development," lies between the justice and environment points and, as Campbell (1996) notes, is the most elusive. Traditional economic development has not succeeded in achieving a balance between justice and the environment. Successful resolution of this conflict is one in which equity is increased and the environment protected in either growing or steady-state economies. Campbell (1996) observes the following:

> Poor urban communities are often forced to make the no-win choice between economic survival and environmental quality, as when the only economic opportunities are offered by incinerators, toxic waste sites, landfills, and other noxious land uses that most neighborhoods can afford to oppose and do without. (p. 299)

Robbins, a South Chicago inner-ring suburb and one of the poorest minority communities in the nation, illustrates the trade-offs poor communities are forced to make. Beginning in 1985, Mayor Irene Brody, presented with no other development options, began trying to attract a waste incinerator. In 1987, Reading Energy, based in Philadelphia, agreed to construct one in the village. Despite considerable local protest over potential environmental damage, Brody saw it as the only hope for creating jobs and economic development. The proposed facility would burn up to 1,600 tons of household garbage a day and was projected to take in 10% of Cook County's trash, turning part of it into electricity. Reading Energy

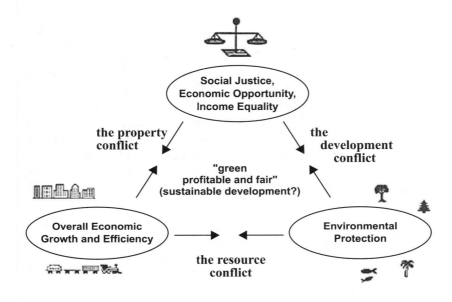

Figure 1.1.   The Development Conflict Triangle
SOURCE: Campbell (1996, fig. p.). Reprinted by permission of the *Journal of the American Planning Association.*

projected it would create 80 permanent jobs. Robbins would own the facility, but it would be operated by the Foster Wheeler Company, which agreed to pay the village a percentage of net revenues, estimated at $1 million a year. This money was to be used for economic development projects in the area. The facility was economically feasible only because of the 1987 Illinois Retail Rate Law, which allowed incinerator operators to write off a large share of their operating costs, making Reading Energy eligible for $385 million in interest-free state loans over 20 years. After years of negotiation, Reading began construction in 1994.

The plan began to unravel in 1996 when Illinois repealed the Retail Rate Law as there was no longer a perceived landfill crisis. Despite the loss in subsidies, construction continued. The subsidy loss meant that Foster Wheeler's revenues would diminish, leaving less economic development money for Robbins. Further, the village would not receive property taxes from the facility, because it had designated the 585-acre site to house the facility, a tax increment finance district. The $438 million facility began operations in June 1997 with 75 employees. The U.S. Environmental Protection Agency cited the Robbins plant with 779 violations during its first seven months of operation. The plant was not viable without the subsidy and closed in October 2000.

Robbins is a sad illustration of not being able to resolve Campbell's three development conflicts. Vast public resources were spent on a facility that brought environmental damage, debt, and an unattractive abandoned facility as a sore reminder. Campbell argues that only by simultaneously resolving the development and property conflicts can economic growth be more fairly distributed and poor communities be able to restore and protect their environments.

Ignoring these two conflicts throughout the postwar period, and particularly during the most recent upswings in the business cycle, resulted in widely acknowledged growing income inequality at the same time that the unemployment rate fell to an all-time low. Despite the unprecedented economic expansions in the U.S. economy in the 1990s, many inner cities and inner-ring suburbs have not shared in the growth resurgence. The movement back to downtowns, and the creation of new middle- and higher-income housing within them, often through conversion of former office and industrial space, has been highly profiled. However, this has not benefited the poorest neighborhoods (the inner city as opposed to the downtown) and, in fact, may be contributing to the displacement of poor residents.

A positive exception can be found in Chattanooga's (TN) efforts to resolve the development conflict. After being labeled the dirtiest city in America by the U.S. Environmental Protection Agency in 1969, Chattanooga has transformed itself from a declining manufacturing city to one with a more diverse and environmentally sustainable economic base. In the early 1970s, the city created the Air Pollution Board to monitor industrial pollution. By 1989, after spending $40 million on pollution control, the city was in full compliance with federal air quality standards.

The process of redevelopment is perhaps as notable as the accomplishments. In 1984, Vision 2000, a planning body with representation from all of the city's neighborhoods and constituencies, was created to develop a new economic and community development agenda for the city. This process resulted in a plan for development of the downtown and riverfront, new economic activity, and inner-city housing development. A river walk and aquarium became the catalyst for downtown development. One pollution control strategy, a 17-vehicle electric shuttle bus system, has turned into an economic development success. Local Advanced Vehicle Systems, the manufacturer, is exporting the buses to at least 14 U.S. cities and also to other countries. Although all economic development projects have not been successful, such as a zero-emissions eco-industrial park that did not prove financially feasible, new clusters of economic activity in the insurance and candy and snack industries are growing. Even though Chattanooga's landscape is still marred by abandoned industrial wasteland, there has been a sustained effort to resolve development conflicts. More recently, the city launched the Chattanooga Regional Growth Initiative, and it is working with Michael Porter's Initiative for a Competitive Inner City to identify and expand key economic clusters.

Campbell's (1996) challenge "to deal with the conflicts between competing interests by discovering and implementing complementary uses" (p. 300) is at the heart of the challenge for economic development planners. As the Chattanooga strategy reveals, economic development planning is the ideal umbrella—particularly in a market-driven economy— under which to resolve these conflicts.

Phase 5: Privatization and Interdependence

Our current phase of economic development is characterized by two approaches to economic development: one that relies on market solutions and another that promotes metropolitan or regional strategies. As the prosperity of the 1990s reduced unemployment rates and welfare rolls, attention shifted to market solutions to revitalize inner-city neighborhoods. The Initiative for a Competitive Inner City (ICIC),[4] created in 1994 by Michael Porter of Harvard University's Business School, has received considerable attention for its emphasis on entrepreneurship and competitiveness as the key to revitalization. Porter founded this national, not-for-profit organization to work in selected cities to overcome misperceptions and promote private sector investment focused on meeting unmet local demand and creating clusters of economic activity that integrate with the regional economy. The good times of the 1990s also meant that urban development was not a national priority. The lone federal program focusing on inner-city economic development, the Empowerment Zone (EZ) program, uses public incentives to create new businesses and markets in poor inner-city neighborhoods. At the same time, an expanding movement, the "new regionalism," which includes the new smart-growth movement, is calling for metropolitan or regional strategies to prevent the decline of inner cities and inner-ring suburbs while stemming urban sprawl.

Porter's market-driven approach to inner-city redevelopment has gained followers in several cities. He argues that inner-city neighborhoods have competitive advantages that have been overlooked, including strategic location, integration with regional clusters, unmet local demand, and human resources. By locating in these communities, Porter believes, the private sector can revitalize them and also turn a profit. He argues that the appropriate role of government is to foster a healthy business climate in inner cities by assembling and improving sites, preventing crime, upgrading infrastructure, providing subsidies, and streamlining regulation. Although acknowledging that many federal programs to cure inner-city ills have been misguided, several urban planners point out the naïveté of Porter's suggestion that there is little role for government in inner-city economic development (Ross & Leigh, 2000). In fact, the role of government in the private sector movement to inner cities is more central than Porter admits. Government facilitation and financing of private investment, particularly in creating public-private partnerships that integrate minority firms into the mainstream market, has been instrumental in local economic development (see Bates, 1997; Harrison & Glasmeier, 1997). Further, simply attracting businesses to low-income areas will not result in income redistribution, because most of them draw labor from outside the area and pay wages too low to support families. Rather, business development efforts must be supported by government provision of adequate housing, schools, and day care (Fainstein & Gray, 1995). Porter would not argue with this role for government and thus appears to be sitting on both sides of the fence.

The Empowerment Zone program is the only federal initiative to promote development in inner cities enacted under Clinton's two consecutive presidential terms during the 1990s. Empowerment Zones (EZs), established in 1994, combine physical revitalization and business development with education, training, and human services within a defined geographic area (Schwartz, 1994). The first six urban areas to have EZs designated were Atlanta, Baltimore, Chicago, Detroit, New York, and Philadelphia. Camden, Los Angeles, and Cleveland were added in 1998, and an additional 15 EZs were added in 1999. Each EZ receives $100 million over 10 years.[5] Each of the designated areas has poverty rates averaging 20% or more and at least 35% of residents with incomes below the poverty level. The goal of the EZ program is to create self-sufficiency by creating employment opportunities for zone residents. Businesses are attracted with tax incentives, performance grants, and loans for locating inside and hiring residents of an EZ. Employers in the EZs are eligible for wage tax credits, worth up to $3,000 (20% of the first $15,000 in wages) for every employee hired who lives in the EZ boundaries. After residents are employed, the goal is that the communities will become revitalized places where stable families flourish.

A U.S. Department of Housing and Urban Development (HUD) evaluation (1999b) of the program claimed the EZs and Enterprise Communities had created 20,000 new jobs and attracted $4 billion in private investment. University of Michigan Planning Professor Margaret Dewar points out that HUD's analysis assumes that all development is the result of EZ incentives. She concludes that the HUD study and U.S. General Accounting Office statistics on business use of incentives in the EZs are unable to demonstrate that the new economic activity would not have happened anyway. Examining employer behavior in three business districts in the Detroit EZ, Dewar (2000) found that the incentives did not affect investments, expansions, or location decisions. Only one of the districts hired local residents.

Other critics of the EZ program suggest that the resources allocated are inadequate and not properly focused. Although $100 million may seem to be a significant investment, it only amounts to approximately $50 per zone resident per year (see Goozner, 1998). Jonathan Levine (1994), a transportation planning professor at the University of Michigan, has observed that $100 million is only enough to pay for 2.1 miles of the suburban highways built in the Detroit area. Porter (1997) argues that public resources such as those funding the EZ program are targeted toward meeting residents' immediate needs rather than generating jobs and economic opportunity that will mitigate the long-term need for social programs. A final critique of EZs and other "development in-place" strategies is that they are rooted in a separate but equal philosophy: "quarantine 'them' in inner city ghettos and barrios away from 'us' and help 'them' build from within" (Rusk, 1993, p. 121).

The other major issue emerging in the current phase of economic development is that of urban sprawl, the unabated spread of development outside the more densely populated urban core. Arguments for stemming sprawl are rooted in concerns about environmental degradation, decreased quality of life, increasing inequality, decreased government efficiency, and reduced economic competitiveness (see Greenstein & Wiewel, 2000). In response, many urban analysts have identified strategies to achieve more balanced city-suburban growth (see Orfield, 1997; Hughes, 2000; Persky & Wiewel, 2000). Most of these focus on ways of achieving regional cooperation and address issues such as metropolitan government, suburban affordable housing, growth management, reverse commuting, impact fees, or tax-based sharing. Others examine assumptions that drive sprawl, such as that distant suburbs are the "least cost" location for many businesses. Studies of business movements to suburbs reveal that privately optimal decisions of firms may raise the total costs to society imposed by continued movement of development from cities to suburbs and back again (see Persky & Wiewel, 2000). Profit alone does not ensure economic efficiency, and certainly not equity or sustainability.

In the early 1990s, several urban scholars began empirically documenting the interconnections between central cities and their surrounding suburbs. Some authors identified income disparity between cities and their suburbs as a cause of decline in overall metropolitan growth. Several studies found correlations between suburban and inner-city income changes and concluded that city-suburb income disparity causes decline in metropolitan growth while more equal income distributions causes growth (see Rusk, 1993; Savitch, Collins, Sanders, & Markham, 1993; Barnes & Ledebur, 1995). These studies sparked considerable methodological debate, particularly over the assumption that correlation implied causality.[6] However, few argued with the basic conclusion that the economies of cities and their suburbs were interrelated.

Some local and state government officials and planning practitioners have taken an aggressive stance to promote more balanced city-suburban growth. In 1973, the state of Oregon began requiring that all cities develop a comprehensive economic development plan and develop growth boundaries appropriate to accommodate anticipated growth over a 20-year period. Cities must identify in their plans how they will create the infrastructure needed to support development (Geddes, 1997). Evidence suggests this requirement has helped to stem sprawl. While cities such as New York and Chicago experienced 65% and 46%

increases in urbanized land with only 8% and 4% population increases, the corresponding figures for Portland are 14% and 11%.

At a metropolitan level, Cleveland has spawned the first government-led advocacy organization in the nation to focus on ending state and federal policies that subsidize suburban sprawl (and, thereby, new suburban development). Observing the same policies that "sucked the life out of urban centers" and are hurting older suburbs, the First Tier Consortium wants to change public investment priorities so that funds for transportation, housing, education, and job development go to existing downtowns and neighborhoods rather than to subsidizing urban fringe development (Schneider, 1999). Further, they advocate a tax-based sharing plan that would have wealthier suburbs contributing to redevelopment of abandoned downtown areas (Garland & Galuszka, 1997).

One reason that metropolitan government and sprawl containment initiatives are practiced in so few cities is that sprawl is not simply an economic or spatial phenomenon, but one with political, social, and racial motivations. Movement to increasingly distant suburbs is partially rooted in the desire to be separated from the racial other (Rusk, 1993, 2000; Powell, 2000). This reality makes it difficult to gain support for metropolitan initiatives from both white suburbanites and minority inner-city residents. The appeal of metropolitan solutions in this context is that they promote mobility for those locked in poor inner-city neighborhoods. Indeed, several critics point out that "in-place" strategies have never been able to ameliorate concentrated poverty (see Foster-Bey, 1997; Powell, 2000). Nevertheless, many in minority communities believe that metropolitan solutions, particularly those that seek to disperse minority populations, will reduce what little political power they have and create cultural dilution (see Powell, 2000). At the same time, many white suburbanites resist metropolitan-wide solutions because they do not see their fates as being tied to the inner city and do not want to share resources through tax-based sharing or expanded school districts.

The Twin Cities in Minnesota and Portland, Oregon, illustrate that metropolitan strategies can make limited, but real, inroads into segregation and concentrated poverty. Metropolitan planning has been present in the Twin Cities region since 1967 with the creation of the Metropolitan Council. In the 1970s, it was well-known for creating the nation's most effective, suburban affordable housing programs. During the 1980s, the Metropolitan Council became less active under a Republican governor (Orfield, 1997). The state's Livable Communities Act, passed in 1995, reinvigorated metropolitan planning. The act created funding for affordable housing, pollution cleanup, and transit-oriented economic development. Minneapolis and St. Paul are among the cities that have signed on to this voluntary program. Cities must meet benchmarks in the three areas to receive funds. The combined Council, Livable Communities Act, and tax-based sharing have had some effect on creating development opportunities in low-income communities. Tax-based sharing has reduced the fiscal disparity between the most affluent suburbs and cities from 50:1 to 12:1 (Orfield, 1997). Likewise, Portland's Urban Growth Boundary has had some impact in reducing racial and economic polarization in the region. However, Portland has a relatively low minority population. When the growth boundaries were enacted, the black population was mostly concentrated in the northeastern section of the city. Portland State University Planning Professor Deborah Howe notes that blacks are being dispersed to inner-ring suburbs because whites are buying up lower-priced housing in this area. Thus, reduced minority concentration has been the result of displacement. The outcome seems to be an unintended, and perhaps undesired, consequence of planning. As Mier (1994) noted, key to the success of metropolitan strategies is honest discussion of racial issues and specific measures to address them. These discussions, unfortunately, are not a high priority in the current phase of economic development.

What Economic Development Should Be

This brief history does not indicate a steady progression of greater acceptance of a more equitable and sustainable approach to development. Examples of such approaches do exist, but they are rare. However, a more sustainable and equitable approach to development is often feasible. What precludes it is mostly the constellation of political forces. In this section, we address the practical challenge of carrying out a development strategy that reconciles equity and sustainability objectives with traditional goals of economic growth and job creation.

First, however, we need to distinguish between growth and development. Economic development is more than just creating growth, wealth, or jobs (see Fitzgerald & Meyer, 1986). It can even be argued that there are times when wealth, job creation, or both should not be the desired outcome in economic development planning if, for example, in the process, income inequality increases, the environment is irrevocably harmed, marginalized groups are made worse off, or all three. Thus, economic development is concerned not only with creating growth or stabilization, but also with specific criteria for distributing the benefits of that growth (or redistributing it within the stabilization). Whereas economic *growth* is defined as more development, more jobs, more taxes, and so on, we define economic *development* as raising standards of living and improving the quality of life through a process that specifically lessens inequalities in metropolitan development and the metropolitan population's standard of living. Further, our distinction between growth and development is not oriented solely to the present because economic *development* is *sustainable*: It is growth and change that neither contributes to rising inequalities nor diminishes opportunities for future generations.

We know that we present a tall order to local economic development planners and practitioners with these criteria for practice. Very little of what is being practiced, despite good intentions, qualifies. However, without a specific articulation of what economic development should be, there cannot be a conscious commitment to modify planning and practice to an economic outcome that creates a more just and sustainable life for all.

This focus on equity in economic development has yet to be officially adopted by the largest membership organization representing practicing economic developers, the American Economic Development Council (AEDC). The AEDC (1998) defines economic development as "The process of creating wealth through the mobilization of human, financial, capital, physical and natural resources to generate marketable goods and services" (p. 18). While AEDC has an 11-point Code of Ethics, only one of the points addresses socioeconomic issues: "Members will consider long-term economic sustainability, environmental issues, and the impact on natural and human resources in dealing with specific projects, policies, or procedures" (*Point 10 of the "Code of Ethics,"* 1998).

Those engaged in economic development practice within the planning field *specifically* are given direction from their larger membership organization to pursue more equitable outcomes. The Code of Ethics and Professional Conduct of the American Institute of Certified Planners statement that, "A planner must strive to expand choice and opportunity for all persons, recognizing a special responsibility to plan for the needs of disadvantaged groups and persons, and must urge the alteration of policies, institutions and decisions which oppose such needs," indirectly, but unmistakably, acknowledges equity as a desired goal. In 1994, the American Planning Association (APA) expanded the goals of comprehensive planning to include community equity, which it defines as expanding opportunities for betterment to be made available to those communities who need them most, and creating more choices for those who have limited options. This responsibility, according to the APA, means it is the obligation of planners to reduce inequalities and narrow the gap opened by disparities in the distribution of resources.

We think that more guidance is needed on how equity and sustainability can be incorporated into economic development practice and offer three overarching principles to that end. Although these principles provide a framework for the goals of local-level economic development planning, individual planners cannot be held to achieving them. Acknowledging the organizational constraints under which planners work, the principles offer individual, local economic development planners direction for how, when, and where they can make choices in formulating their communities' strategies.

The three macro principles are the following:[7]

Economic development should increase standards of living. In the United States, the minimum standard of living is defined as that which lies just above the federally established poverty level. The poverty level is the minimum amount of income, adjusted by household size, necessary to provide a life-sustaining standard of living. The minimum standard of living, therefore, is relative to or defined in comparison with other groups (e.g., those falling just below the current poverty level, or those making two times the poverty level). Both the poverty level and the minimum standard of living have risen over time. They do so alongside the moderate (or middle) and high standards of living. From a consumer basket approach, a parallel argument for the minimum standard of living could be made to the middle standard of living:

> What represented the middle standard in the 1960s would not constitute the middle standard in the 1990s. Items such as microwave ovens, camcorders, and VCRs, which did not exist in the 1960s, have become staples in middle-class homes of today. The home within the middle-class consumption basket has changed as well; two bathrooms and a two-car garage are becoming the norm. (Leigh, 1994, p. 32)

The preceding focuses on consumer goods, but there are other items (services and benefits) we associate with a rising standard of living, for example, increasing numbers of households receiving paid health care benefits and retirement plans, taking vacations, and sending their children to college. Most often, the rise in standard of living that provides these consumer goods and services is achieved through the growth of better jobs that economic *development* creates.

In local economic development, recognition of the need for a minimum standard of living is often represented by calls for job creation that provides "living wages," that is, earnings high enough to lift an individual or family above the poverty level (Harrison & Bluestone, 1988). One of the most distressing trends in job creation in the final two decades of the 20th century has been the proliferation of low-wage, no-benefit jobs that do not lift workers out of poverty. As a consequence, the ranks of the working poor have grown (Theodore, 1995; Edin & Lein, 1997; Wright & Dwyer, 1999). Thus, the need for local economic developers to facilitate the creation of more living-wage jobs becomes ever more pressing.

While acknowledging that local planners have very limited ability to alter the structure of employment, specific strategies aimed at increasing the quality of jobs include the following:

- Technical assistance to local businesses to adopt high-performance production systems

- Sectoral workforce development initiatives that foster the development of high-wage, high-skill industries (see Chapter 2)

- Coherent linking of education and job training with economic development ensuring that more people have the skills needed for living-wage jobs (see Chapter 7)

Economic development should reduce inequality. A goal of economic development as practiced by any city or town should be to reduce inequality among different demographic groups (age, gender, race, and ethnicity) as well as spatially defined groups such as indigenous populations versus in-migrants or old-timers versus newcomers. At a macro or regional level, these practices can work toward reducing inequality among different kinds of economic or political units (small towns vs. large cities, the inner city and suburbs, rural areas and urbanized areas, underdeveloped and developed regions and nations).

Inequality manifests itself spatially. Rapid metropolitan growth focused on the suburban edge and based on in-migration of skilled professionals from other parts of the nation has been a force for growing urban inequality. It typically means that the original residents experience higher costs of living, as well as widening gaps in quality of life and economic opportunity between the metro region's inner city and inner ring and its newer suburbs. In many cities, the location of new job development in the outer-ring suburbs has created a spatial mismatch between employment opportunities and jobless or underemployed residents of inner cities.

The spatial mismatch is also racial. Economist Joseph Persky uses a representation index to examine the share of suburban employment held by African American and Latino workers. His analysis reveals that minority workers, particularly blacks, have not benefited proportionally from suburban service-sector job growth. This differential is even more evident in better-paying service jobs. In general, whites have greater ability to move closer to their places of employment. Blacks, because of racial segregation in housing, have more limited options (Massey & Denton, 1993). Whites working in the suburbs are more likely to live there, whereas African Americans and Latinos are more likely to commute from the inner city.

Strategies for inner-city revitalization, and increasingly the inner-ring suburbs that are home to America's new immigrants, must take into explicit account the influence of racism. As Rob Mier (1994) observed, "Race is a powerful aspect of most planning situations in urban areas, yet it too often is the last way a problem, or especially an opportunity, is framed" (pp. 235–236). That inner cities are not treated as part of regional economies and, instead, are treated as separate, independent economies—that is, ghettoized—can largely be attributed to racism:

> Suburbanites not only fear traveling into the inner city but also having inner-city residents traveling into their communities. This factor plays a part in suburban communities' refusal to allow city mass transit lines to extend to the suburbs . . . [For example] In the Atlanta city-region, the two highest growth counties of the metropolitan region, Gwinett and Cobb, can not get enough workers to fill many of their retail and other service sector positions. As a consequence, jobs such as those in fast-food establishments, which normally pay at or close to minimum wage are being bid up to attract suburban teen and retired workers. Both of these counties, whose population is approximately 90% white, have refused to allow the Metropolitan Atlanta Rapid Transit System (MARTA) to be extended to them. (Ross & Leigh, 2000, pp. 376–377)

We argue that unless this racism is addressed explicitly, inner cities and inner-ring suburbs will never be an integral part of their metropolitan regional economies. Thus, reducing inequality, in both a spatial and racial sense, requires both "in-place" and mobility strategies, including more coherent metropolitan-wide planning. The same four economic development strategies suggested previously to increase earnings can be applied to reducing inequality.

Economic development should promote and encourage sustainable resource use and production. With regard to goods consumption, a rising standard of living attained through sustainable resource use and production requires a very different approach than that which has yielded our previous advances. It requires recycling the material goods cast off by an increasingly affluent and consumer-oriented society, which in turn can spawn economic development activities based on ecological-industrial principles. It also requires greater controls on growth to stem greenfield consumption and sprawl proliferation. Stemming sprawl requires planning strategies aimed at reusing previously developed properties and vacant urban land. In this sense, economic development practice must be closely linked to land use planning (Leigh, 1999, 2000).

Recycling and energy conservation can create economic development opportunities based on ecological-industrial principles. In doing so, this kind of economic development activity can lessen dependence on landfills, prevent further environmental degradation, and reduce demand for raw and nonrenewable materials and waste associated with primary processing. These recycling activities can be sited in eco-industrial parks that are located on inner-city or inner-ring suburban brownfield or obsolete industrial lands (Leigh & Realff, 2001). Eco-industrial parks have been promoted by the President's Council on Sustainable Development as models for industrial efficiency, cooperation, and environmental responsibility that represent unique opportunities for communities to create jobs and protect the environment in a manner reflective of their values (President's Council on Sustainable Development, 1999).

In a closed-loop eco-industrial park, the waste for one industrial process or system is used as the inputs or enabling resource of another (Ehrenfeld & Gertler 1997; Dunn & Steinemann 1998; Dwortzan, 1998). The most well-known and successful eco-industrial park is Kalundborg, Denmark. Begun in the 1970s by a core group of partners seeking to reduce costs and satisfy regulatory requirements, the park has evolved into a firm network that includes an oil refinery, power station, pasteboard factory, pharmaceutical plant, sulfuric acid producer, and a range of agricultural and aquacultural development activities. These firms engage in a complex exchange of materials and energy, an industrial symbiosis, enabled by infrastructure innovations, which create a contiguous connection between them. As Dunn and Steinemann (1998) observed, "As of 1994, the \$60 million investment in infrastructure had produced an estimated \$120 million in gross revenues and cost savings for the partners in the symbiosis" (p. 666). The environmental benefits of the Kalundborg development include reductions in consumption of oil, coal, and water, as well as atmospheric emissions of CO_2 and SO_2 and increased recycling of fly ash, sulfur, gypsum, and nitrogen sludge.

To date, there are no eco-industrial parks in the United States that match the integration of Kalundborg. More typical of those established in this country is the Civano Industrial Eco Park of Tucson, Arizona, which brings together businesses with certain core capabilities (e.g., makers of PVS electric vehicles, circuit boards, steel fabricators, design firms, etc.) to share resources and participate in joint operations such as water treatment and transportation (President's Council on Sustainable Development, 1999).

The eco-industrial park, in its most complete, symbiotic form, is the ideal model of sustainable resource use and production. It is also a model that is in its infancy and requires much further technological and industrial organizational advances before it can be widely adapted. In the meantime, a number of the economic development strategies we focus on in this book can help to meet, at least partially, the principle of sustainable resource use and production. They include the following:

- Brownfield redevelopment (see Chapter 3)

- Industrial and office property reuse (see Chapter 6)

- Industrial retention (see Chapter 4)

- Commercial revitalization (see Chapter 5)

These are also strategies that could be used as complements to eco-industrial parks.

Sustainability with development requires creating better options for poor (and most often minority) communities by looking past short-term goals of simple job creation and increasing tax revenues. Quick or desperate economic growth fixes such as allowing the siting of a hazardous waste incinerator in a poor urban neighborhood preclude successful long-term economic development by reducing the possibility of attracting quality commercial and residential development.

Using these principles, we can define local economic development: Local economic development preserves and raises the community's standard of living through a process of human and physical infrastructure development based on principles of equity and sustainability.

Incorporating Equity and Sustainability Into Local Economic Development Practice

We offer this definition while acknowledging that many, if not the majority, of economic development planners and practitioners hold jobs that require them to simply implement existing economic development programs or address the growth goal exclusively. Economic development planners perform a multitude of tasks, depending on where they practice. Those working in smaller towns define their jobs almost exclusively as marketing and industrial attraction. Industrial attraction dominates for those working in the economic development arms of utility companies and chambers of commerce. Those working in the planning departments of large cities have a wider range of responsibilities, including small business development, industrial retention, negotiating tax-increment financing districts (TIFs), export promotion, commercial revitalization, workforce development, and brownfield redevelopment. Essentially, most economic development planners market, make deals, implement programs, or all three.

This reality raises a critical question as to the political feasibility of implementing economic development as we have defined it. Is a vision of economic development planning that includes growth, equity, and sustainability only attainable in rare historic moments when progressive governments are in power? (For example, Cleveland in the 1970s, or Chicago during the Washington administration of the 1980s.) We answer this question in two ways. First, any given economic development program has to be analyzed in the context of an overall strategy or vision. Within any long-term strategy, trade-offs and compromises have to be made along the way. We have talked about the Washington administration in Chicago promoting social justice by balancing the interests of the neighborhoods against wealthy downtown developers and increasing participation in the decision-making process. Yet the Washington administration also negotiated a deal to build a new baseball stadium, Comiskey Park, with considerable state and city subsidy and led the negotiations for a similar deal for the Chicago Bears that fell through. These can hardly be considered deals that advanced any of the three principles.

When the owners of the Chicago White Sox announced that they were looking for a new location for Comiskey Park, city planners explored rebuilding the stadium across the street from the stadium, the only workable location for the park in the South Armour Square neighborhood.[8] This site was home to numerous small businesses and homes owned primarily by low-income African Americans. The residents formed the South Armour Square Neighborhood Association to protest the plan, which had the support of local politicians, team officials, and high-ranking major league baseball officials. Economic Development Commissioner Robert Mier tried to appease both sides.

In June 1988, the legislature approved construction of a new stadium with $150 million coming from state bonds and the rest from a 2% hotel tax (Fort, 1997). In addition, the White Sox would only have to pay rent at the new park if attendance surpassed 1.2 million a year (Cagan & deMause, 1998). Owners of the 178 homes and 12 community businesses dislocated received market value for their property plus a $25,000 cash bonus and moving expenses as compensation for being displaced (Betancur, Leachman, Miller, Walker, & Wright, 1995). The neighborhood association also secured an air-conditioning system for a public school threatened by dust from the construction. The residents of the affected public housing complex, Wentworth Gardens, received no compensation.[9]

Although the City was able to extract a few benefits for the poor African American community, the Comiskey Park deal could hardly be viewed as progressive politics given that the $167 million stadium opened in 1991 was 100% financed with city and state dollars. Yet the political reality was that if Washington were to get reelected, he could not be the mayor who lost the White Sox or the Bears. The solution was to extract as many benefits as possible to save the larger economic development agenda of achieving balanced growth and creating employment opportunities that would raise incomes in poor neighborhoods. Analyzing this effort without understanding the overall economic development strategy would lead one to a very different conclusion.

A second response to the question of the feasibility of our approach is that the same strategy can be implemented in ways that are responsive to issues of equity and sustainability or not. In each of the following chapters, we present cases of some of the most commonly used economic development strategies and demonstrate how they can build toward greater equity, sustainability, or both. Further, we examine their potential in inner-ring suburbs as well as city settings. We summarize each chapter next.

Chicago looms large in our examples, simply because it is a city in which there is a long history of economic development planning. Because of Chicago's economic diversity, these efforts covered a number of areas and often represented the most innovative practices in the country. Further, because of Chicago's racial and ethnic diversity and a history of engaged community politics, many of the political issues in economic development are illustrated in practice here. Most of the Chicago practices we discuss are applicable to inner-ring suburbs or smaller towns.

Chapter 2 on sectoral strategies reveals some of the trade-offs that have to be made when deciding whether to create employment opportunities for less or more highly educated residents. Two cases are presented: a sectoral initiative focusing on manufacturing in Chicago and one focusing on biotechnology in New Haven. We raise the question of how effectively a city can maintain and attract traditional and high-tech industries. Some are questioning whether Chicago has missed important opportunities for developing new industries or promoting modernization of existing sectors. Critics of new industry attraction campaigns acknowledge the importance of developing new industries but argue that they affect only a small percentage of residents. The point is that both are important, and city governments have to figure out how to do both to create opportunities for increasing incomes.

Chapter 3 on brownfield development focuses on the major new challenge for local economic developers that emerged during the decade of the 1990s but will remain with us far into the future. The challenge of redeveloping contaminated and underused land (most often commercial or industrial) is greatest for local economic developers practicing in central cities and inner-ring suburbs. This chapter provides these local economic developers with an understanding of the issues that brownfields pose for promoting sustainable economic development. It identifies the range of barriers that thwart efforts to clean up and redevelop brownfields, as well as the current gaps in resources and approaches for promoting equitable brownfield redevelopment. It teaches planners how to characterize the extent of a community's brownfield problem and provides examples of communities that have developed programs for brownfield redevelopment at the local level.

In Chapter 4 on industrial retention, we detail the political battles that occurred over establishing planned manufacturing districts in Chicago. The day-to-day activities of city planners in implementing this strategy involved studying land use patterns, refiguring zoning maps, meeting with various stakeholders such as businesses and community organizations, attending public meetings, political organizing, and attending City Council meetings. The overarching goal of this innovative planning tool was to keep well-paying jobs in the city—a step toward raising overall standards of living. Further, planned manufacturing districts promote sustainable development because they create an alternative to developing new greenfield manufacturing space when land already in use for manufacturing is economically viable. Implementation in Portland and Seattle is considerable as well.

Chapter 5 on commercial revitalization focuses on three strategies for fostering a strong retail base for sustainable economic development and improved assets, shopping, tax base, and employment opportunities in a community. First, the chapter examines the tried-and-true principles of Main Street Revitalization for their applicability to aging suburbs and inner cities alike. Second, it examines the unstoppable "Big Box" retail trend and how aging suburbs and inner cities can accept these retail outlets in a manner that minimizes their negative impacts. This chapter identifies the revised zoning and planning codes, as well as design guidelines that some communities have adopted to ensure that "Big Box" retail is a positive addition to their retail base. The third strategy the chapter focuses on is a response to what is the most difficult retail development challenge of all—redoing the growing number of obsolete strip malls and retail strip developments in inner-ring suburbs and inner cities. Profiled in this section are the efforts of a nonprofit group in an old, unincorporated suburb of metropolitan Atlanta to revitalize and redesign a particularly unattractive and underused retail strip on one of the region's most congested roads. The unincorporated status of this suburb significantly increased the difficulty of obtaining resources and support for its revitalization efforts, but 20 years of effort have finally begun to pay off for this suburb named Sandy Springs.

Chapter 6 on the reuse of office and industrial property identifies the trends in office and industrial markets that are contributing to high vacancy rates and underuse of existing urban infrastructure and resources. Technological developments in the office, industrial, and warehousing sectors have changed location and building specification requirements such that many inner-ring suburb and central-city properties have become obsolete for the functions for which the were originally built. This chapter provides local economic development planners with an understanding of these technological developments and other economic forces that are changing the demand for older office and industrial properties. It explores renovation possibilities and limitations for older properties and profiles a number of examples of successful conversions and reuses of these properties from around the country.

Chapter 7 on workforce development discusses the efforts of the Seattle Jobs Initiative to prepare low-income residents for living-wage jobs with advancement potential. On a day-to-day planning level, the initiative has changed the practices of organizations involved in workforce development to be more responsive to the needs of poor residents. As a result, these residents have access to better-paying jobs they would not have otherwise. To the extent this has been achieved, the first two principles have been realized. They have had little success, however, in convincing employers to create more jobs with living wages, benefits, and advancement potential.

Notes

1. The five phases build on a more abbreviated discussion of three phases in Mier and Fitzgerald (1991).

2. Utility companies have been active players in local economic development because it was in their interest to expand their customer base. Adding industrial customers ensures additional residential customers. Because utility companies historically were not mobile, they had proscribed local utility markets, and their growth depended on growth of the local economy. The advent of utility deregulation is lessening utility companies' focus on local economic development in some areas.

3. ITEP is a national research and education organization working on government taxation and spending policy issues.

4. More information on ICIC can be obtained from its Web site, at www.icic.org.

5. An additional 66 cities were designated as Enterprise Communities. They are eligible for up to $3 million in federal funds. In 1998, two Enterprise Communities were moved up to EZ status.

6. As with any new area of empirical inquiry, critiques of the methodology of the early studies emerged. Criticism of the interdependence studies and Rusk's elasticity hypothesis focused on the methodology employed, correlation analysis, arguing that causation could not be inferred from correlation. Hill, Wolman, and Ford (1995) acknowledged that central city and suburban economies are interconnected but argued against the contention that inner-city decline, in itself, leads to suburban decline.

7. The original articulation of these principles can be found in Leigh-Preston (1985.)

8. Soon after Jerry Reinsdorf and Edward Einhorn purchased the Chicago White Sox in 1981, they announced that they were considering several suburban Chicago sites. On the advice of then Governor Jim Thompson, Reinsdorf threatened to leave Chicago for St. Petersberg, Florida, where an empty stadium awaited them. As Governor Thompson put it, if Reinsdorf wanted a new ballpark in Chicago, he would have to convince people that he was willing to move the team (Cagan & deMause, 1998).

9. The residents remaining in the neighborhood, including many who lived in public housing, filed a civil action lawsuit against the newly formed sports authority, the State of Illinois, and the City of Chicago. They alleged that the new stadium site was selected in violation of their civil rights and that an alternate location north of the one where it was built would have displaced fewer businesses and homes. The residents claimed that the site was not considered because it would have displaced white residents of the Bridgeport neighborhood (Betancur et al., 1995). The residents lost the case.

2

Sectoral Strategies for Local Economic Development

Sectoral strategies identify industries or occupations that are a significant source of employment in a local economy and create an environment in which they can grow. The strength of this approach is it focuses economic development efforts on stable or growth industries, allowing economic development practitioners to consider all aspects of an industry's health such as land and facilities, technology adoption, financing, workforce development, marketing, infrastructure, and transportation. As an approach that increases standards of living by creating well-paying jobs, sectoral strategies are receiving considerable attention in both economic and workforce development circles. Indeed, in a field where practice is defined as "Shoot anything that flies, claim anything that falls" (Rubin, 1988), strategies focused on a clearly identified target are the exception.

The problem, ironically, is that the concept is ill defined. A wide variety of initiatives are considered sectoral strategies. The states of California and Connecticut, among others, have identified target industries around which they are building comprehensive economic development strategies. Public-private partnerships in New York City and Los Angeles are revitalizing the garment industry. City government in San Francisco is seeking to create a biotech industry. A community organization in Chicago offers technical assistance and job training to metalworking firms. All these are considered sectoral initiatives, yet they vary significantly in intent, scale, and scope.

The terms *sector, cluster,* and *targeted* industry strategy are often used interchangeably. A sector usually refers to a group of firms that produce similar products, but it can also refer to shared markets, technology, resources, or workforce needs (Okagaki, Palmer, & Mayer, 1998). Clusters are defined by Michael Porter's (1997) Initiative for a Competitive Inner City as "concentrations of companies and industries in a geographic region that are interconnected by the markets they serve and the products they produce, as well as by the suppliers, trade associations and educational institutions with which they interact" (para. 5). Targeted industry strategies are typically more narrowly focused than cluster strategies. Cluster strategies typically are defined by geographic proximity and include several sectors. For example, a cluster targeting the meatpacking industry would also include hog farmers, corn growers, animal feed companies, packaging businesses, waste haulers, and farm equipment dealers (see Mt. Auburn Associates, 1995).

In this chapter, we examine the range of economic development activities defined as *sectoral strategies* and analyze their goals and outcomes. We begin by creating a framework for differentiating among them. We then discuss the limited evidence on how well these strategies work. Then we present case studies of implementation of two distinctly different types of sectoral strategies, both of which offer lessons for economic development practitioners on the extent to which they can expand existing and create new sectors in a city or regional economy. The first is the Jane Addams Resource Corporation, a community development corporation in Chicago leading a sectoral initiative in metalworking. The second is the biotechnology strategy of New Haven, Connecticut.

Overview of Sectoral Strategies

We found it useful to distinguish between sectoral strategies that target existing industries in an economy and take measures to improve their productivity and create conditions for their growth ("Type 1") and those that identify new growth industries that could be attracted or created to diversify the local economy and create more high-skill, high-wage jobs ("Type 2"). The goals of each may be different. In some cases, a targeted existing industry is not experiencing employment growth. Some manufacturing industries, for example, are stagnant or declining in employment, but still have many job openings due to replacement. In contrast, sectoral strategies in some high-tech sectors create few jobs but aim to create strengths in industries with high growth potential, positive externalities, or high multipliers.

A sectoral strategy is not necessarily a new program, but rather a way of integrating existing programs. Active participation by employers is essential, as it is only through their cooperation that economies of scale and scope can be created. In most sectoral strategies, employers are encouraged to work together in technology adoption, employment training, and research and development (Mt. Auburn Associates, 1995). Dresser and Rogers (1998) point out three efficiencies achieved through sectoral strategies:

- Economies of scale are created by working with several firms with shared labor force and other needs.

- Economies of scope are leveraged when workforce development programs crosscut the specific occupations of individual firms in an industry.

- Positive network externalities are created when firms work cooperatively to solve common problems. (p. 72)

Both types of sectoral strategies can be undertaken by public-private partnerships at any level of government or by nongovernment organizations (Table 2.1). Usually, one organization acts as an intermediary that coordinates all aspects of the initiative. An intermediary assesses the needs of businesses in the targeted sector and the extent to which existing programs meet them. Recommendations are then made to state and local government to add needed services. Intermediaries make links for businesses in the sector to government and community services and the labor market. Thus, the success of sectoral initiatives is highly dependent on the entrepreneurial ability of the intermediary. Many types of organizations can play the intermediary role, including city or state government agencies, community colleges, universities, community-based organizations, chambers of commerce, unions, or industry associations. Although the sectoral initiatives in Table 2.1 are classified by the lead intermediary, in all cases partnerships between state and local government, community organizations, and employers are present in varying degrees.

TABLE 2.1 Typology of Sectoral Strategies

Type/Level of Sectoral Initiative	State	Metropolitan	City	Community
Type 1 Maintain/ Build Strength of Existing Industry	In 1984, the Massachusetts Executive Offices of Economic Affairs and Labor earmarked funds for Industry Action Projects. The projects examined trends in key industries in specific geographic regions of the state. Effective projects were developed in the needle trades and other areas of manufacturing. The Machine Action Project (MAP), in the Springfield area, created workforce development and technical assistance programs in metalworking and printing and graphic arts. MAP also identified plastics conversions technology as a potential growth industry. The programs were unique in the level of union participation. Funding for the action projects was eliminated after Governor Weld took office.[a]	The Wisconsin Regional Training Partnership (WRTP), organized in 1992, advocates for living-wage jobs in manufacturing in the Milwaukee region. This consortium of business, labor, and community leaders has almost 50 member firms (with combined employment of 50,000 machining, electronics, plastics, and related sectors). The WRTP initiatives offer modernization, lifetime learning for incumbent workers, disadvantaged job-seeker training and placement, and school-to-work transition. The WRTP partners with the Milwaukee Jobs Initiative and the Central City Workers' Center (Milwaukee's One-Stop Employment Center) for placing central city residents in manufacturing jobs.	The Garment Industry Development Corporation (GIDC; NYC) was organized in the late 1980s to support New York City's garment industry, the largest manufacturing sector in the city. The nonprofit organization involves labor, industry, and government in providing job training, technical assistance, marketing and export development assistance, and assistance in finding local providers of goods and services. GIDC operates out of three centers in different parts of the city. Industry representatives serve on the board and fund training through a collective bargaining agreement in which employers contribute to a union-managed training program.[b]	The Jane Addams Resource Corporation (JARC), a nonprofit community development corporation in Chicago, provides technical assistance and training to employees of metalworking firms and training for un/underemployed workers. Organized in 1985, JARC has moved from working with individual firms to state and national industry trade associations in delivering job training and identifying career ladders in metal-working trades. In collaboration with community organizations and government agencies, JARC is a partner in two advocacy organizations that are seeking to influence state policy on incumbent worker training.

(Continued)

Table 2.1 Continued

Type/Level of Sectoral Initiative	State	Metropolitan	City	Community
Type 2 Attract/Create New Industry	In 1998, **Connecticut's Economic Competitiveness Strategy** identified six industry areas that are key to the state's economy. The industries are a mix of traditional and high tech: tourism, financial services, telecommunications, health services, manufacturing, and high technologies. Clusters have been identified within each area. Each cluster team has identified action steps to attract new or modernize existing industries and to provide workforce training, capital, and infrastructure in the identified industry.	In 1998, **the San Diego Association of Governments** (SANDAG) identified 16 clusters as drivers for the regional economy and targeted the 9 high-tech clusters as the focus of economic development activities. SANDAG's regional plan elaborates several activities to promote growth, including attracting venture capital, creating workforce development partnerships, developing a land inventory and incentives to promote redevelopment and infill development, expanding the base of affordable housing, reducing regulatory burdens, and developing a strategy for low-level radioactive and other hazardous wastes.	**Tucson, Arizona**, has identified environmental technology (products and services that eliminate, prevent, reduce, or remediate environmental problems) as one of six clusters. The initiative is led by a business membership organization, the Environmental Technology Industry Cluster (ETIC). The Greater Tucson Economic Council supports ETIC by hosting an annual trade show that has increased sales of environmental technology equipment manufacturers and recruits new businesses. Working with the Arizona Department of Commerce, ETIC has conducted foreign market studies to provide information to members on how to expand exports.	N/A

SOURCE: a. Fitzerald and MacGregor (1993).
b. Conway (1999).

In the nonprofit and foundation world, *sectoral strategy* has come to mean strategies that focus on creating employment opportunities for low-income people in a specific sector or cluster of occupations (see Clark & Dawson, 1995). The definition developed by the Center for Community Change (Okagaki et al., 1998) is widely used: Sectoral strategies target a particular occupation within an industry and intervene by becoming a valued actor within industries that employ the occupation. Similarly, Dresser and Rogers (1998) define sectoral strategies as development of intermediaries that provide solutions to industry problems and use those solutions to improve training for incumbent workers and increase access to the industry to disadvantaged workers. The goals of sectoral strategies, then, are to assist low-income people to obtain well-paying jobs and to create systemic change within that occupation's labor market. The Aspen Institute (Clark & Dawson, 1995) identifies goals this type of sectoral strategy can achieve:

- Establish a higher standard of wages and benefits for employees in an occupation

- Change the hiring standards or hiring practices of employers

- Leverage greater employer investment in skills training

- Establish better defined career ladders for occupations

- Create new educational and training programs at community colleges and vocational/technical institutes

- Promote collaborative approaches among groups of employers and supporting institutions (p. 7)

For the most part, sectoral strategies targeting low-income populations focus on strengthening existing industries. The key intermediary in these strategies is often a community-based organization. Workforce development is the key program activity in all the initiatives focused on existing industries in Table 2.1.

State and city governments are most likely to be the key intermediary in "Type 2" strategies. This is because the level of economic analysis needed to identify target industries or sectors and the level of financial resources is beyond the capacity of most community organizations. Our search uncovered little involvement by community organizations in

"Type 2" strategies, particularly those focused on high-tech industries. In government-led sectoral strategies, public services provided to the private sector typically include financing, affordable land, technology transfer, management assistance, and walking companies through regulations. In addition to helping individual firms, the goal is to develop networks and institutional connections among customers, suppliers, competitors, government, labor, and trade associations. Workforce development is sometimes, but not usually, a central component of the strategy. The goals relate to industry growth, regional exports, and general job growth.

Many states are moving to "third wave" economic development strategies that build institutional and human capacity to create a competitive environment (Eisinger, 1995), particularly for high-tech industries. In these strategies, state government takes on the entrepreneurial role of starting new enterprises, technology, and products, and identifying new markets for state businesses. Pennsylvania's Ben Franklin Partnership, created in 1982, is one of the most successful programs in promoting entrepreneurship in advanced technology industries. Networks comprising universities, government, and entrepreneurs in the four quadrants of the state apply for challenge grants for new product development. The grants are evaluated based on the likelihood of bringing the product to market, job creation, and ability to attract private sector investment. A 1999 evaluation found the program generated 21,800 jobs in grant-receiving firms between 1989 and 1996, at salaries 45% higher than the state's average (Nexus Associates, 1999). Considering jobs and tax income, the study estimated the state received a 14-to-1 return on investment (Nexus Associates, 1999).

There is a general shift to state programs that focus on attracting high-skill jobs in key clusters or sectors in each region of the state (Waits & Howard, 1996). Programs often include promoting networking, technology transfer, and new business development (Bradshaw & Blakely, 1999). Yet many states have economic development plans that include a combination of strengthening existing industries and creating new ones. The plans identify key sectors in regions of the state on which economic development efforts are focused. The Industry Action Projects in Massachusetts (Table 2.1) have given way to the Massachusetts Technology Collaborative, created in 1995 to foster the growth of the state's high-tech sector. California's first statewide economic development plan, produced in 1993, divided the state into six regions and targeted two economic clusters in each (see Kleiman, 2000). Oregon's focus on key high-tech sectors contributed to transforming a lagging lumber-based economy into a thriving high-tech one. In each case, easily solvable problems in specific industries were identified and solutions provided.

Cities are also targeting new high-tech industries as they find themselves losing out to suburbs in attracting these jobs. High-tech jobs account for 25% of new employment in cities, but the rate of growth in suburbs is 30% faster than in cities (U.S. Department of Housing and Urban Development, 2000). These initiatives have a strong emphasis on finding land for development. Land banking, brownfield redevelopment or infill development are key components of the high-tech sectoral strategies of New Haven (presented later), San Diego, and San Francisco (Table 2.1).

The Planning Process for Sectoral Strategies

Targeting requires in-depth knowledge of the industry. Six separate, but interrelated, aspects of the industry are typically considered (see Wiewel & Siegel, 1990; Theodore & Carlson, 1998):

1. *National and regional context:* National and regional trends in the targeted sector are suggestive of how the sector could perform locally. Some industries that are not growing nationally may still be candidates for targeted industries due to local specialization or need for replacement workers. Most targeted industries, however, are growing at the national and regional level.

2. *Technological change:* In some industries, the rate of techno-logical change moves too fast for many firms to keep abreast of them. Small- and medium-sized enterprises, in particular, need assistance in when and how to adopt new technologies. In addi-tion, marketing advice and management assistance are often provided.

3. *Location issues:* Many industries, including those as diverse as manufacturing and biotechnology, are locating in distant sub-urbs. Sectoral strategies require understanding the factors behind firm location decisions to maintain existing or attract new firms. Factors may include availability of specialized labor pools and access to local supplier industries.

4. *Labor force needs:* The availability of skilled labor is a pressing need for firms in almost every sector. The location of a skilled labor force is one of the most important location considerations

firms in many industries consider. A survey of sectoral strategies by Clark and Dawson (1995) found job training to be the most frequently used programmatic tool, although training was not typically targeted to low-income populations.

5. *Regulation:* Helping firms respond to environmental regulation is an important economic development function. In biotechnology firms, assistance in meeting federal regulations in disposing of hazardous waste is an important service. In addition, firms need assistance in meeting state and local permitting requirements.

6. *Financing:* Part of the strategic-planning process is identifying sources of financing for business startups or expansions. Some sectors (e.g., high-tech) are more in need of venture capital, whereas others need loans for bricks and mortar.

Sectoral strategies do not necessarily require creating new programs. Many states and cities have small-business assistance programs, loan pools or venture capital funds, one-stop permitting and licensing offices, and technology assistance programs that can tailor services to the needs of targeted sectors. Local Workforce Investment Boards have been put in place to coordinate job placement and training services as mandated by the Workforce Investment Act of 1998 (see Chapter 7). These services also can be tailored to the needs of specific sectors in the local economy.

Do Sectoral Strategies Work?

Even researchers who have developed methodologies for identifying industries for targeting express doubts about the ability of the public sector to pick winners (see Carlson & Wiewel, 1991). In the case of older industries in which there is a political commitment to workers, the choice is less problematic. It may be more difficult to achieve consensus on new growth industries to target, especially given weaknesses in the data.

In a forum on targeted industry strategies in *Economic Development Quarterly*, Buss (1999) points out several flaws with the data and the reasoning behind targeted industry strategies. First, the ES202[1] SIC codes cannot keep up with new industries, particularly the high-tech, high-growth industries that are of interest in targeted industry strategies. Second, inferences about local economies are derived from national data, when in fact how an industry is doing in any given region or city may bear little relation to the industry nationally. Buss (1999) identifies numerous cases in which the industries identified for targeting are meaningless and concludes random selection of an industry is as likely to pick a winner as the most sophisticated techniques available. Third, Buss argues that targeted industry strategies inefficiently attract economic activity to places other than least-cost locations. Finally, Buss suggests that the most damning evidence against targeted industry strategies is the absence of rigorous evaluations or even convincing circumstantial evidence.

In response, Wiewel (1999) points out that Buss offers no empirical evidence of flaws in studies used to choose target industries and that there are few economic benefits to these strategies. Wiewel argues if perfect data were required, few economic development initiatives would be undertaken. Finkle (1999) suggests Buss's inefficiency argument assumes the private sector has perfect information and always identifies its most efficient location, an assumption that is falsified by the number of business failures. In fact, there is a whole literature in economic geography that suggests many firms, particularly those in high-tech industries, can create their own factors of production in many different settings (see Storper & Walker, 1989).

Buss is indeed correct on the lack of solid evidence on the effectiveness of targeted industry strategies, but more rigorous evaluations are under way. None of the sectoral initiatives in the Mt. Auburn study had developed performance measures or a process for monitoring impacts. Recognizing the lack of reliable data on the outcomes of sectoral initiatives, the Aspen Institute launched the Sectoral Employment Development Learning Project (SEDLP) in 1996 to document and evaluate the outcomes of six sectoral employment programs throughout the country (see Table 2.2).

The study focuses on workforce development outcomes. To date, baseline characteristics, and intermediate outcomes 12 months after training are available. Preliminary evidence shows SEDLP participants earn more than Job Training Partnership Act (JTPA)[2] participants but still do not earn enough to raise a household over the poverty line (Conway, 1999). As the study progresses, more detailed and longer-term outcome data will become available.

Even when focusing on one aspect of a sectoral strategy—workforce development—evaluation criteria are not clear. One of the problems each of the organizations in the Aspen study faces is they receive several sources of funding, each with separate evaluation and reporting requirements.

TABLE 2.2   Aspen Institute Sectoral Employment Development Learning
Project (SEDLP) Sites

Site	Targeted Industry or Occupation
Bronx: Cooperative Home Care Associates	Home health aides, certified nursing assistants
Chicago: Jane Addams Resource Corp.	Machine tooling and metalworking
Detroit: Focus: HOPE	Machine tooling and metalworking
New York: Garment Industry Dev. Corp.	Garment industry
San Antonio: Project QUEST	Health care, financial services, environmental technology
San Francisco: Asian Neighborhood Design	Specialty furniture and wood manufacturing

SOURCE: Compiled by the author from the Aspen Institute Web site: www.aspeninst.org/
cop/eop_sedlp.html

In sectoral initiatives that link workforce development, land acquisition, financing, and other services, it is hard to identify which factors
contribute to business and job growth. Further, some of the benefits
of sectoral strategies are difficult to measure. Outcomes of joint
marketing efforts are often measured by activities such as participation
in trade events, but there is little information available on money
saved. Other outcomes, such as creating networks, have no agreed-on
measures.

In presenting the two following cases, our intent is not to evaluate
the efforts but to identify the steps involved in developing and implementing sectoral strategies and the actors and intermediaries involved, to
compare their outcomes to the goals identified previously, and to offer
suggestions on implementation.

The Cases: Sectoral Strategies
in a Traditional and High-Tech Sector

Jane Addams Resource Corporation

Jane Addams Resource Corporation (JARC), a nonprofit community development corporation organized in 1985, was created to
preserve and strengthen the industrial base of the Ravenswood community on Chicago's Northwest side. The community contained a healthy

urban manufacturing sector until the 1980s when manufacturers began migrating to the suburbs. As manufacturers moved, the area began to gentrify and pressures to convert manufacturing spaces to residential lofts ensued, threatening the 200 manufacturers still in the community. JARC's success in maintaining the area as a vibrant manufacturing center is due to its role as an intermediary in helping firms stay competitive and providing effective training programs in the metalworking trades. Further, JARC advocates for state and local economic development policy to maintain and expand the viability of manufacturing in urban neighborhoods.

The first step in JARC's planning process was to conduct a needs assessment of local metalworking firms. The study revealed the fragility of the small metalworking firms due to the loss of many large equipment manufacturers for which they were suppliers. The survey identified a need for assistance in adopting new technologies and in problem solving to reduce costs and increase productivity. Further, it identified a need for skills upgrading that could provide basic skills and technical training for incumbent metalworking employees.

As a Local Industrial Retention Initiative group (see Chapter 4), JARC began providing technical assistance to metalworking firms and initiated several programs to help local firms become more competitive, such as coordinating joint bidding on projects or orders that none could do independently. Larger Ravenswood firms agreed to subcontract with smaller firms, creating vertical networks. JARC also facilitated networking among the firms through a monthly peer learning and support group and works with firms in upgrading their technology and production processes. JARC encourages firms that have invested in new technologies to share their experiences with other firms and holds seminars on topics such as Occupational Safety and Health Administration (OSHA) compliance.

The next step was to develop the training programs. JARC began the Metalworking Skills Training Program in collaboration with four metalworking companies and Truman College, one of Chicago's city colleges. The Metalworking Skills Program offers several courses for incumbent workers that provide 20 and 48 hours of training in 6- to 8-week modules.[3] The pilot program started in early 1991 with seven workers and currently serves 280 workers and 30 companies per year. By 2000, JARC had trained more than 1,200 workers in at least 70 companies. Since 1997, the program has had a completion rate of over 90%.

JARC staff work closely with employers in developing curriculum, using tools such as the DACUM (Develop A CurriculUM) process. This job analysis tool uses expert workers in focus groups to develop competency-based curriculum and then confirms the validity of the data with managers. A team of shop floor workers and management from 18 local companies worked with JARC staff on developing the curriculum for several of the courses in 1994.

An ongoing task is to obtain funding for training programs and the supporting labor market research. An ongoing funder is the Illinois Prairie State 2000 Authority, a state agency that provides grants and loans to small and medium-sized manufacturers for retraining workers. Increasingly, JARC relies on fees collected from employers. Companies are required to pay the workers for attending 16 of the 32 class hours and to pay fees to cover the match requirement on some of the more expensive courses. Over time, employers have become more involved in terms of both paying for training and participating in curriculum development.

As demand for the Metalworking Skills Program grew through the early 1990s, JARC raised funds from foundations and employers for a training center. Over several years, JARC purchased three buildings in the area and housed the training program and its administrative offices in one of them.[4] Local metalworking firms invested in equipment and instructors. By 1996, the facility was in full operation as the only die-setting apprenticeship program registered with the U.S. Department of Labor. Demand for courses continues to grow, including firms located in Cook, Lake, and DuPage counties near Chicago. Despite the Metalworking Program in place, JARC could not meet the demand of local firms for skilled employees. The Opportunities in Metalworking Program (OMW) was created to provide unemployed and working poor residents with pre-employment skills and vocational training. The 7-week program uses the metalworking training staff as instructors in mathematics, blueprint reading, and basic metalworking machinery operation and setup. Local firms provide plant tours and conduct mock interviews with students. Job placement and postplacement assistance is provided and participants are encouraged to enroll in other courses. Through OMW, JARC is demonstrating to employers that poor urban residents can become skilled and reliable employees.

Early in the development of the Metalworking Skills Training Program, JARC staff became interested in identifying career progression ladders for low-skill workers. JARC secured a grant from the Joyce Foundation in 1994 to identify career ladders in metalworking occupations and to develop curriculum leading to career advancement. A follow-up grant in 1996 was used to develop the Metalworking Skills Assessment Tool, a series of validated tests to screen and evaluate employees in five areas: basic math, applied math, applied measuring, blueprint reading, and manufacturing technology. As described by Anita Flores, JARC's associate director, "This tool is unique because the modular form of the scoring allows the user to develop an individualized training plan for each test taker by subject area" (personal interview, June 2001). It took 2 years (1997–1999) to develop and validate the tool. It is now being marketed in partnership with the Precision Metalforming Association.

The Metalworking Skills Assessment Tool is being used with JARC's curriculum to place students in the appropriate class based on the set of competencies identified for each position in the career ladder. As students take courses and gain job experience, they move through three levels of punch press operating, two levels of punch press die setting, and lead die-setting positions. Punch press operators start at $6.50 per hour and can advance to $12.00 with training and experience. Die setters start at $7.00 per hour and can advance to $17.00.[5]

JARC has also been active in encouraging the state to support its approach to linking economic and workforce development. In 1996, JARC became a coordinating partner of the State Agenda for Community Economic Development (SACED). Due to SACED's efforts in educating legislators on the need for community-based employment training, Illinois passed the Job Training and Economic Development Demonstration Grant Program in August 1997 (Public Law 90-0474), which awarded $1 million of new funds to 18 demonstration programs. Regulations for the program have been written, and proposals have been solicited.

In October 2000, JARC launched the Incumbent Workers Policy Advocacy Subcommittee. The group is developing a policy agenda to present to state legislators to create targeted training resources for those identified as the working poor and to extend wage supports and benefits to this group. Machine operators in the Chicago metropolitan area earn between $6.00 and $20.00 per hour, with a median of $11.64 (U.S. Department of Labor, Bureau of Labor Statistics, 2000). Clearly, access to training allows many to move from working-poor to living-wage status.

JARC's success as an intermediary is attributable to several factors. In its role as a Local Industrial Retention Initiative (LIRI) group, JARC staff members are in touch with local firms and understand their employee skill needs. JARC integrates education and training into its economic development work by translating its good relationships with local employers into training and jobs for work-needy people. Because local employers trust JARC, they are willing to work with staff in developing curriculum and providing instructors. JARC is now working with industry associations as well as individual firms. In the policy arena, JARC has been able to connect with a very active network of community-based organizations that advocate for policies to create advancement opportunities for low-wage workers.

JARC stacks up well against the six goals of sectoral strategies identified by the Center for Community Change (Okagaki et al., 1998). The training programs are well respected by employers. The Metalworking Skills Assessment Tool has given JARC further legitimacy with employers, evidenced by the involvement of state and regional manufacturing associations. JARC has engaged employers in identifying career ladders for workers, thus establishing a higher standard of wages and benefits for employees in metalworking occupations. Because more firms are using this sophisticated assessment tool, JARC has changed their hiring and advancement practices. Further, many employers are now willing to invest more in worker training. JARC has promoted collaborative approaches among individual employers, employer associations, city government, and other community organizations involved in economic development.

New Haven, Connecticut: Just in Time in Biotechnology

Biotechnology, the application of biological knowledge and techniques to develop products and services, is the focus of many "Type 2" sectoral strategies. Specifically, biotechnology uses modern recombinant DNA or monoclonal antibody methods to manufacture biological products.[6] It is one of the most research-intensive industries in the world, with the U.S. biotech industry alone spending over $9 billion in R&D annually. The U.S. biotechnology industry currently employs more than 153,000 people in high-wage, high-value jobs in 1,283 companies.[7] As a cluster of industries that creates high-wage jobs and creates many spin-off jobs, biotechnology is a very attractive focus of sectoral strategies (Bowles, 1999).

Biotechnology, like high-tech industries, has location needs that preclude its locating in many places. A major research university is at the core of all biotechnology clusters and is a necessary, but not sufficient, need. Once this requirement is met, state and local economic development policy can facilitate the expansion of biotechnology clusters in many ways. This case study examines the relative significance of Yale University, New Haven, Connecticut, state economic development agencies in creating a biotechnology cluster. Although each played a different role, individual efforts eventually became mutually complementary. We focus on New Haven because Yale, the city, and state were relative latecomers in targeting biotechnology. Although it is often thought the early bird gets the worm in high technology and that latecomers only get the low-value-added segment of the industry, this case suggests otherwise. Surrounded by a thriving biotechnology cluster to the north in Boston and to the south in New Jersey, New Haven's cluster is expanding rapidly.

Yale University's Role in Biotechnology

Yale University is key to the New Haven area's biotechnology cluster. Unlike Massachusetts Institute of Technology (MIT), Stanford, and other universities active in high-tech spin-offs throughout the 1980s, Yale did not commit much effort to helping faculty turn their research into marketable products. Several factors converged to push Yale into technology transfer in the mid-1990s. First, because many other universities were far ahead in this area, it was becoming increasingly difficult to retain and hire new faculty. It is quite common for potential faculty members to ask while interviewing what the university will do with their patents, business ideas, and so on. By the early 1990s, it became apparent that expanding the university's interaction with the private sector was essential to attracting the best and the brightest faculty.

A second factor was Richard Levin's appointment to the presidency of Yale in 1992. An economist, Levin's research interest was in the role of universities in high-tech development. Aware of the role of high-tech startups as a creative outlet for faculty in science and medicine and in maintaining the university's research reputation, Levin reversed the university's long-standing reluctance to pursue commercial ventures. The

Yale Office of Cooperative Research (OCR) would become much more active under Levin. The OCR mission includes oversight for patenting and licensing university inventions and working with Yale researchers to identify inventions that could become commercial products and services useful to the public. An important goal for the OCR is to identify new ideas, cultivate venture funding for them, and facilitate their development into companies that become part of the New Haven economy. Levin directed OCR to step up its activities and hired new staff members to work with faculty and existing biotech companies. In addition, OCR began working in areas normally within the purview of city economic development planners, such as showing entrepreneurs around the area. Thus, in the mid-1990s, Yale's OCR became an important actor in local economic development.

In February 2000, Levin announced Yale would "spend $500 million on university biomedical programs aimed at turning New Haven into a major biomedical center over the next 10 years" (cited in Rosenberg, 2000, p. G1). In fact, the university has committed $1 billion over the next 10 years to expand its basic science, engineering, and clinical research programs. This includes $500 million for new construction and another $500 million on the Science Hill section of campus. In addition, the Yale medical school is building a $176 million research facility.

The university's efforts have worked. In 1999, Yale was the third highest university generator of royalties, at over $46 million. OCR Managing Director Jon Soderstrom (personal interview, October 2000) points out that with an endowment of over $10 billion, royalties are not what motivates Yale's involvement in this arena. Yale's goal is to be competitive in attracting and keeping top faculty and to contribute to the economic health of the New Haven region. Indeed, the region's economy has benefited enormously. In 2000 alone, Yale spin-offs have attracted $1 billion in capital investments. Currently, OCR is a partner in 24 high-technology startups, employing over 1,000.

Connecticut's BioScience Cluster

Yale's efforts are complemented by state and city efforts to support the growth of biotechnology. Connecticut is one of many states targeting biotechnology in its economic development strategy (Biotechnology Industry Organization, 2000).[8] These programs offer various tax credits, startup capital, incubator facilities, technology development centers, and job training. *Connecticut's Economic Competitiveness Strategy* (1998) is organized around Michael Porter's conception of industrial clusters. In 1998, legislation was passed to create the Connecticut Industry Cluster Initiative. Bioscience, which includes biotechnology, pharmaceuticals, and academic research, is a subgroup of the high-tech cluster.[9] Bioscience was targeted because it could build on an existing cluster of pharmaceutical firms in the state and because of high wages and economic spin-offs.

The cluster is overseen by Connecticut United for Research Excellence, Inc. (CURE), a not-for-profit, 501(c) 3 corporation. CURE facilitates information exchange and communication among cluster members and publicizes the cluster's contributions. The state provided the BioScience Cluster with $150,000 in seed money from the Department of Economic and Community Development and biotech firms contributed $350,000. CURE now has over 100 dues-paying members, including bioscience companies, hospitals, professional societies, venture capitalists, and economic development organizations.

As the convening center for the state's bioscience research sector, CURE maintains data on the sector, identifies problems and threats to firms in the sector, and develops strategies to advance the group. Debra Pasquale, CURE's president, relates several examples (personal interview, November 2000). Member companies often complain about the time and expense involved in obtaining state permits. Indeed, CURE, working with the state Office of Policy and Management, identified 220 permits in 20 separate agencies for which new companies have to apply. CURE staff worked with state agencies to reduce the number of permits to 25 and created a link on the CURE Web site to the agencies requiring permits. Another commonality many bioscience companies have is staying in compliance with federal laws governing the use of animals in research, clinical trials, and low-level radioactive materials. CURE organized committees to identify best practices and offers seminars to disseminate these practices. CURE recently organized a job fair in which 16 companies participated. The companies have found by organizing together they attract more potential workers and increase the chances of finding jobs for domestic partners.

CURE encourages BioScience Cluster firms to promote their common interest. In 1996, several firms united to lobby state legislators to create a biotech tax incentive. The incentive exempts biotech companies from sales, use, and property taxes for 5 years and provides a 15-year carry forward of R&D tax credits. The carry forward for net operating losses against future taxable income recognizes the cost time involved in taking a biotech product from R&D through the regulatory approval process. In 1999, the incentives were expanded to allow companies to exchange R&D tax credits for cash and to extend net operating losses forward 20 years (existing legislation allowed 5 years).

The cluster's BioScience Facilities Fund is managed by Connecticut Innovations (CI), a quasi-public agency that is the state's technology investment arm. CI was created in 1993 to supply equity investment to high-tech startups and has become the state's leading public sector investor in high technology.[10] The Biotechnology Facilities Fund was started with $30 million in state bond funds and $10 million from CI. The fund is now self-sustaining through repayment and equity appreciation on its investments. To date, loans of $18.5 million have been committed, and the entire $40 million will be fully committed within a year. CI has invested $16 million to develop 200,000 square feet of laboratory space for eight companies in the New Haven area.

CI provides a small but significant part of the total financing package for any given startup. Soderstrom (personal interview, October 2000) estimates it takes approximately $5 million for a startup. The biggest expense for a new bioscience firm is lab space, which costs up to four times as much per square foot as other office space. Venture capitalists are more likely to fund the research and development component, leaving a gap in bricks-and-mortar financing. By filling this gap, CI plays an important role in facilitating startups.

CI also makes risk capital investments in high-tech companies. Once a company's product is successful or the company goes public, CI withdraws support. Initial investments range from $100,000 to $1,000,000, and the company must be able to raise matching funds, typically from venture capitalists. Companies can receive funding from both funds. In the last 5 years, over $75 million in venture capital has been invested in Yale spin-offs. Since then, several of these companies have gone public. Of them, Alexion has raised $548 million, and Curagen, $200 million in secondary offerings. Both companies are in the drug development phase.

The BioScience Cluster is the fastest-growing cluster among the six. In its *Fifth Annual Economic Report,* CURE reports that between 1995 and 1999, bioscience R&D expenditures in the state increased by 75% (from $1.5 to $2.6 billion) and employment increased by 42% (from 8,455 to 12,020). The main beneficiary of the state's bioscience targeting is New Haven, a city of 126,000.[11]

New Haven's Role in Promoting Bioscience

The city of New Haven also has a role in maintaining and attracting bioscience firms. New Haven Mayor John DeStefano has implemented an aggressive economic development agenda since taking office in 1994. In the early 1990s, many growth sectors in the national economy were declining in New Haven. Wholesale trade, retail trade and finance, insurance, and real estate, for example, experienced declines of 6% to 10% between 1993 and 1997. The unemployment rate in 1994, when the mayor took office, was 8.5%, compared with 6.9% for the state. Clearly, in addition to strengthening existing industries, New Haven needed a new growth industry. Considering the presence of Yale and five major pharmaceutical companies in the area (see Figure 2.1), bioscience was a good choice.[12]

New Haven employs a multipronged strategy for building a bioscience sector. Henry Fernandez, director of Economic Development and Planning, identifies four areas in which the city promotes its growth: property development, zoning and permitting, eliminating red tape in startups and expansions, and improving the city's quality of life.

Although the majority of biotech startups in most parts of the country are in suburbs (U.S. Department of Housing and Urban Development, 2000), significant expansion is taking place within New Haven, much of it because the city has been very active in property development. The city was a partner and owner (with Yale and Olin Corporation) of Science Park, a largely vacant industrial park, incorporated in 1981, with the goal of transforming it into a business incubator for high-tech firms. Science Park is located in an enterprise zone wedged between Yale's campus and Newhallville and Dixwell, two low-income communities in New Haven. In the 1980s, several computer software and related companies located in the park. Some of these succeeded and moved to other locations, and others folded. By the late 1980s, the park was badly in need of tenants. A few bioscience companies located in the park in the early 1990s. In 1998, the Connecticut Housing Finance Authority committed $14 million for demolition of old buildings, renovation, and new construction. Currently, there are 34 companies in the park, including eight bioscience firms and four high-tech research labs or production facilities. Three of the biotech companies came in since the 1998 renovation.

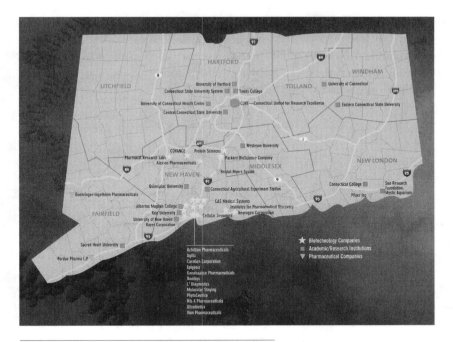

Figure 2.1.   Connecticut's BioScience Clusters
SOURCE: Copyright 2001 by Connecticut United for Research Excellence Inc. (CURE).
    Reprinted with permission.

In the fall of 2000, a private company, Lyme Properties, took over as owner and developer of the park. When fully developed, Science Park will have one million square feet of biotech space, restaurants, business services, parking, and landscaped parks. The park operates like an incubator because it provides shared equipment, facilities, technical assistance, and access to Yale's library. As part of the city's commitment to creating opportunities for people in the surrounding communities, the development plan calls for several buildings to be redeveloped for light industry.

An abandoned Southern New England Telecommunications building has been renovated into laboratory and office space for biotech companies by a private developer. The George Street building (see Figure 2.2) is located in the heart of New Haven, near the Yale medical school. The 500,000-square-foot building was purchased for approximately $20 million, and slightly more than that is being spent on renovation. Although city government could not have made that kind of investment, the building was attractive to the developer because it was wired for fiber optics and had companies waiting to occupy it. Two companies, Achillion Pharmaceuticals (see Box 2.1) and Molecular Staging, opened their laboratories before construction was completed. The building can accommodate category I and II research facilities.

Box 2.1.
Achillion Pharmaceuticals, Inc.

Achillion, started by scientist and entrepreneur William B. Rice, is the largest biotech company startup in the history of Connecticut (and one of the largest in the nation). Started in February 2000, the company focuses on discovering and developing anti-infective and antiviral agents for the treatment of hepatitis B and C and HIV. Achillion will be involved in all four phases of product development: preclinical trials, clinical trials, product development, and commercialization.

Achillion illustrates how existing relationships and active promotion of New Haven landed the company. Rice intended to locate in Princeton, New Jersey. He had assembled impressive scientific and management teams and venture capital to develop an antiviral agent against HIV. At the last minute, the venture capital funder pulled out. At around the same time, Yale researcher Yung-Chi (Tommy) Cheng had developed several new antiviral compounds but did not have a management team in place to start a company to develop and test them. One of Achillion's founders, Jerry Birnbaum (head of preclinical development of antivirals at Bristol Meyers Squibb) knew Cheng and linked him to the company. Rice started exploring a partnership with Yale, but the university's policy is not to work with companies unless they locate in the New Haven area. After touring the area with Buzz Brown of the Yale Medical School, Rice decided to locate in the George Street building.

Achillion started with $17.2 million in financing from several venture capital companies and Connecticut Innovations. The company received an additional $6.5 million from the same investors in November 2000.

Figure 2.2.   The George Street Building in New Haven
SOURCE: Photo provided by John Soderstrom; reprinted with permission from
    Winstanley Enterprises, LLC.

Another million-square-foot science park is being constructed in nearby Hamden (see Figure 2.1), also on a former manufacturing site. The three spaces will create 2.5 million square feet of bioscience laboratory space in the New Haven area. After venture capital, the biggest need of new firms is real estate. A key role for city government is land accumulation (Dostaler, 2000). Anticipating the need to develop more properties, New Haven has provided approximately $400,000 for testing 12 former industrial sites for reuse by small manufacturing and biotech firms, with plans for eight more over the next 2 years. Talks are continually under way with landholders to assemble larger parcels.

Fernandez is eager to develop brownfield sites for two reasons. First, it gives the city more options in how to use the greenfields that are available. One use for which Fernandez is "saving" greenfield sites is for future expansion of the city's theatres or other arts buildings. Further, if he starts developing greenfield sites for biotech firms, it will be harder to sell the older sites. The second reason is to create jobs in the city so low-income residents can benefit from the high-tech boom as well.

An economic development item of which Fernandez is particularly proud is finding a first-rate building inspector. One normally does not think of a building inspector as a critical economic development need, but it is for the biotech industry. Biotech laboratories have zoning specifications with which most building inspectors are not familiar. No other industry has quite the same set of needs for airflow, waste disposal, and so on. The city conducted a national search for a building inspector who thoroughly understood these needs and could work with developers in identifying all the modifications needed to meet federal regulations. The approach of the building inspector and planning staff is to find solutions, even if it means going to the state to request modifications in the code. As planning goes, Fernandez concludes, "It's not very glitzy, but it's what makes the difference in getting a firm up and running quickly and getting bogged down in regulatory delays" (personal interview, October 2000). Another step the city has taken is to move all permitting into one office. These steps are complementing Yale's efforts and making New Haven an attractive place for new startups.

Another indirect way of creating an environment conducive to biotech or any high-tech development is improving the city's quality of life. New Haven's reputation centers on its strengths in science, higher education, and the arts. These elements reinforce each other, and the city is trying to facilitate all three. The city's role, in all areas, is to be the facilitator, not the deal maker.

A weak link in both state and local efforts is workforce development. Soderstrom (personal interview, October 2000) predicts there will be a labor shortage as the number of biotech workers is likely to double over the next 5 years. Although the majority of positions require college degrees, there are many lab positions that only require community college training. Yet until last year Middlesex Community College, in Middletown, had the only associate degree program in the state.[13] When the program started in 1994, it was difficult to convince biotechnology firms that people with only associate degrees could fill their positions. Recently, CuraGen, a New Haven genomics company, tried a few interns and then hired them permanently. CuraGen has grown from 15 to over 300 employees and has ongoing needs for new employees. To promote the program, CuraGen now makes equipment donations and has funded five $1,000 scholarships. Other companies are now hiring Middlesex graduates.

The program graduates 10 to 15 students per year, but does not have the resources to expand, despite demand throughout the state. Neither of the community colleges in New Haven offers a biotechnology program. Given the labor shortage already being experienced by New Haven biotechnology programs, developing a program locally is the minimum first step needed in workforce development.

Lessons for Economic Development Practitioners

The four concerns raised by Buss have not proved to be problematic in either of these cases. The first concern was whether available industry data are inaccurate. Yet the data available in Connecticut were adequate for identifying key sectors, and JARC used a simple survey to identify the workforce development needs of metalworking firms. Buss is correct that relying on national data to identify target sectors locally is likely to produce as many misses as hits. In the case of Connecticut, comparing national growth industries to existing state strengths allowed for identification of both old and new industries that would be likely candidates for growth. National employment in manufacturing is declining, yet JARC staff knew that many firms in Chicago were experiencing labor market shortages due to retirements and lack of awareness of employment opportunities in manufacturing by young people. Finally, both initiatives take documentation and evaluation seriously. As part of the Aspen Institute Sectoral Employment Development Learning Project, JARC is setting standards for how sectoral strategies should be evaluated. CURE is keeping records of firm startups and employment growth through annual surveys.

We categorized JARC as a community initiative and New Haven as a city initiative. Yet both initiatives required cooperation from other levels of government, public sector agencies, and the private sector. JARC draws on state and city funds for its technical assistance and training programs, which are increasingly being subsidized by the private sector. The private sector, in the form of both individual firms and industry associations, plays an increasingly active role in promoting the growth of the metalworking industry. One could summarize the New Haven case by simply saying "but for Yale," none of it would have happened. Yet state support in providing gap funding and promoting networking, and the city's efforts in creating an infrastructure to support startups have been essential in attracting and keeping firms in New Haven.

The focus on manufacturing in Chicago and bioscience in New Haven reveal trade-offs inherent to sectoral strategies. A city can choose several target sectors, but conflict can emerge over which is allocated more resources. As discussed in Chapter 4, Chicago planners have had to choose between two seemingly incompatible goals: maintaining or attracting middle-class residents and maintaining businesses that provide jobs for working-class residents. A *Business Week* (Weber, Gogoi, Palmer, & Crockett, 2000) cover story suggests Chicago has focused too much on old-economy businesses and thus has not been able to capitalize on the high-tech industries of the new economy. The article quotes several leading Chicago businesspeople and academics who claim that Chicago has missed out on high tech. Among Chicago's missed opportunities due to lack of venture capital are Marc Andreessen, a University of Illinois student who moved to California to start Netscape Communications Corporation, and Lawrence Ellison, who moved to Silicon Valley to create Oracle Corporation, the second largest software company in the world. Further, the article points out that local universities are not promoting faculty startups. It may indeed be true Chicago focuses too much on its existing strengths and not enough on attracting new sectors.

On the other hand, high-tech sectoral strategies are criticized for benefiting only a narrow set of occupations. A recent report by the Center on Policy Initiatives (CPI; Marcelli, Baru, & Cohen, 2000) suggests targeting high-tech development leaves out many residents and may increase inequality within regions. The criticism is based on the San Diego Association of Governments' (1998) cluster strategy that identified 16 clusters as drivers for the regional economy and targeted nine high-tech clusters as the focus of economic development activities.[14] The authors of the CPI report point out that the targeted clusters comprise only 18% of the region's jobs and the number of working poor within them increased throughout the 1990s, with 22% of targeted cluster jobs paying less than $18,000 annually.[15] Further, the targeted sectors are less likely to employ women and nonwhites (partly due to geographic separation).

This critique raises the question of whether "Type 1" sectoral strategies are inherently low wage, low skill, while "Type 2" strategies create mostly high-wage, high-skill jobs. In fact, the goal of JARC and other "Type 1" sectoral initiatives throughout the country is to increase opportunities for wage progression and career advancement for low-wage and entry-level workers. When linked with modernization assistance, sectoral initiatives promoting traditional industries open up a medium-wage, medium-skill development path. Although biotech and other high-tech sectors employ much higher percentages of college graduates, there are significant opportunities for people with associate degrees.

Further, high-tech initiatives can identify complementary employment opportunities and include workforce development for positions requiring less education. Light manufacturing space is being developed at Science Park to create employment opportunities for people in the adjacent low-income community. It remains to be seen how many such jobs are created, and in the meantime, New Haven's poverty rate is 19.5%, significantly higher than the rest of the state (8.3%) or the national average (11.8%).

It appears Connecticut's sectoral initiatives are replacing a corporate subsidy orientation to economic development, but the state has a long way to go in this regard. The three state agencies that provide economic development subsidies to firms have given away $622 million to 1,050 companies since 1991 (Breslow, 2000). Breslow reveals most of the subsidies have not been put to good use: two thirds of the companies subsidized did not meet their employment projects resulting in producing less than one half of the projected employment. In addition, the average subsidy per job, at $54,271, exceeded the federal government limit of $35,000 for economic development programs. The annual subsidy total peaked in 1993 at $153 million and has fallen each year to $47 million in 1999 (Breslow, 2000).

Workforce development is the principal component of JARC's efforts, although it plays a minimal role in New Haven's and Connecticut's bioscience strategy. Currently, JARC is working with employers to build career ladders that provide better-quality workers for employers and better-paying and more secure jobs for residents. The bioscience initiative in Connecticut has only begun to consider workforce development needs. As a longer-term strategy, CURE is promoting science education to primary and secondary school students. Connecticut is behind many states in community college programs in biotechnology. The state needs to provide incentives for more community colleges to develop programs geared to high tech generally and biotechnology in particular.

Both strategies involve the reuse of urban land. JARC's efforts have resulted in many new businesses locating in previously empty manufacturing sites. The three buildings JARC bought and redeveloped house manufacturing and other types of businesses. A unique aspect of New Haven is that the high-tech strategy is combined with a focus on brownfield redevelopment. As discussed in Chapter 3, brownfields are a serious barrier to development in many inner cities and innerring suburbs. Indeed, most high-tech development is in suburbs. While many states and cities are building totally new facilities for biotechnology, New Haven has housed new labs in existing buildings, several in abandoned manufacturing facilities. According to Jon Soderstrom (personal interview, October 2000), part of the reason is although it is usually more expensive to retrofit existing buildings than to build new lab space, it can be done more quickly. In New Haven, there are so many companies in desperate need of startup space that time has become more important than cost. New Haven's Planning Director, Henry Fernandez, has a slightly different take on the reasons. The city is encouraging firms to locate in Science Park to distribute the benefits of high-tech growth to low-income residents and to preserve a small supply of greenfield sites for other types of development.

A final observation is on the role of cooperative leadership in both sectoral initiatives. As an intermediary, JARC has taken organizational leadership with employers in promoting career ladders as good business practice. In both SACED and the Incumbent Workers Policy Advocacy Subcommittee, JARC has taken a leadership role in ensuring that economic development benefits workers in lower-skill occupations. It was the leadership of Richard Levin, president of Yale University, that changed the possibilities for bioscience in New Haven, but once the ball was rolling, leadership in promoting and supporting the industry became a collaborative effort among the university, the state, the city, and the private sector.

Both cases reveal the importance of building relationships in economic development practice. JARC did not have relationships with employers when its work began but developed them over time. A level of trust had to develop before employers were willing to make major investments of time and resources into upgrading jobs. Soderstrom (personal interview, October 2000) points out he did not have relationships with venture capitalists or entrepreneurs in bioscience when he started his position as managing director of OCR. Therefore, he started attending conferences and events attended by venture capitalists and bioscience researchers: "Relationships just don't happen, you have to be purposeful and strategic in making them happen" (personal interview, October 2000). Relationship building is a process that takes time but is essential to the success of sectoral strategies.

Notes

1. The ES202 data are based on reports of employment by industry that states submit to the Department of Labor for Unemployment Insurance purposes.

2. The federal Job Training Partnership Act is the major provider of job training for low-income populations and displaced workers. It has been replaced by other programs under the Workforce Investment Act of 1998 (see Chapter 1).

3. The Metalworking Skills I class combines math literacy with vocational education by teaching shop math at the same time as blueprint reading and use of measuring tools. In Metalworking Skills II, workers move on to any of three areas: Advanced Blueprint Analysis, Measuring for Quality Control or Trig for the Trades. A third level of training provides workers with hands-on instruction in machine training: the operation and setup of a punch press. In addition, JARC offers Business Software, Manufacturing Software (CAD, CAM, and CNC) and English as a Second Language classes.

4. The properties are another component of JARC's economic development strategy. The buildings are retained for industrial use to retain good jobs in the community and to prevent gentrification.

5. These figures were provided by Israel Martinez, Director of Training at JARC.

6. There are many distinctions within this broad definition. Connecticut United for Research Excellence defines biotechnology as activities related to research, product development, or manufacturing of biologically active molecules; devices that employ or affect biological processes; or devices and software for production or management of specific biological information.

7. Most biotech companies are small. Approximately one third of biotech companies employ fewer than 50 employees. More than two thirds employ fewer than 135 people.

8. See www.bio.org/govt/survey.html for a list of state economic development activities in biotechnology.

9. The others are tourism, financial services, health services, telecommunications and information, and manufacturing.

10. CI was spun off from the Connecticut Product Development Corporations, a state-funded quasi-public organization that invests in product development.

11. New Haven's population peaked in 1950 at 164,000. It is expected to decline to 119,500 by 2003 (Connecticut Department of Economic and Community Development, 2000).

12. The current growth environment is one in which existing companies feel secure in making new investments. In 2000, Pfizer opened a 585,000-square-foot

research facility and is building another 400,000-square-foot building in New London. Bayer's pharmaceutical division just completed an expansion of its operations in Woodbridge (see Figure 2.1) with a 33,000-square-foot organic chemistry facility and a 125,000-square-foot R&D facility in West Haven. Boehringer Ingelheim Pharmaceuticals, Inc. announced expansion plans that include a 27,000-square-foot Lead Discovery Technologies Building. This and other buildings under construction will create approximately 250 new jobs (Dostaler, 2000).

13. The program is in environmental science with a biotechnology option.

14. The industry clusters were divided into three categories: high-tech targeted, low-tech nontargeted, and nonclustered industries.

15. Part of this decline is attributable to the rate of unionization in the targeted clusters falling from 14% to 3% during the 1990s. Eight of the 12 clusters experienced declines in weekly earnings inequality due to stagnant earnings at the bottom and rising earnings at the top of the distribution.

3

The Brownfield
Redevelopment Challenge

Introduction to the Brownfield Problem

Brownfields pose one of the most significant forces for unleveling the economic development playing field between, on the one hand, central cities and older suburbs and, on the other hand, young suburbs, edge cities, and the exurban fringe. Defined as previously used parcels of land where knowledge, or merely the suspicion, of contamination hinders future redevelopment potential, the label "brownfields" distinguishes such sites from never-before-developed sites, or, greenfields (U.S. Environmental Protection Agency [EPA], 1995; Bartsch, 1996; Leigh, 1996).

Knowledge and suspicion of a site's contamination typically results in its no longer being considered for redevelopment and can even taint prospects for contiguous sites. This is due to the liability that owners of a contaminated site assume by law, whether or not they were the actual contaminators, and the costs—which may not be fully calculable at the outset—incurred to clean up the site for reuse. The resulting "environmental redlining" or "brownlining" (i.e., identifying areas to be excluded from redevelopment considerations) could significantly dim the economic prospects for the population residing near the site, as well as hurt the municipality's tax base (Leigh, 1994).

Heavy industrial facilities, such as steel mills, as well as chemical and other types of processing plants, are among the best-recognized sources of environmental contamination. However, brownfields can result from ubiquitous service-oriented facilities such as gas stations, auto repair shops, and dry cleaners, which have been prime contributors to the pollution of land, water, and air. Furthermore, the U.S. Environmental Protection Agency (EPA) has estimated that one fourth of the nation's several million underground storage tanks holding petroleum or other hazardous substances for industrial, commercial, and residential uses could be leaking, thereby contaminating the surrounding soil and groundwater. Also contributing to pervasive contamination problems is the historic practice of off-site disposal of industrial waste in landfill sites not designed to contain hazardous and toxic chemicals.

Economic developers should be aware that the burden that inner-ring suburban brownfields place on local governments might exceed that of central cities. First, the resources available to a suburban government to remedy a large industrial site's contamination are unlikely to match those that a large city can put together. Second, suburban areas' greater reliance on well water, instead of municipal water systems, as well as their greater proximity to landfills for industrial waste disposal, make them more susceptible to problems of groundwater contamination (Leigh, 1994).

Because the history of industrialization and urbanization is closely connected with the nation's central cities and first-ring suburbs, they have been left with a legacy of environmentally contaminated sites that can potentially endanger humans' and other species' health, as well as degrade land and water tables. The Comprehensive Environmental Response, Compensation and Liability Act (CERCLA) of 1980 marked the beginning of an official, federal response to environmental contamination. This legislation was specifically established to address the problems associated with land contamination. It assigned responsibilities for cleanup and provided mechanisms for enforcement. Unfortunately, the environmental legislation that has developed over the last two decades—in an effort to safeguard the health of humans and other species—has had the unintended consequence of significantly increasing the difficulties encountered in redeveloping properties with previous industrial, commercial, or both uses found predominantly, but not exclusively, in central cities and inner-ring suburbs.

Planners, land and economic developers, as well as local elected officials seeking to engage in redevelopment are finding that they do not have a level playing field relative to greenfield sites because of the legal, technical, and financial hurdles that must be overcome on brownfield sites. As a result, central cities and inner-ring suburbs are saddled with yet another force (in addition to poorer schools and services, higher crime and unemployment rates, etc.) that can widen the gap in economic opportunity and quality of life between them and greenfield-based suburban development.

Our previous lack of knowledge about the environmental contamination associated with industrialization has meant those that prospered from previous urban economic growth did so at the expense of future generations inhabiting central cities and their inner-ring suburbs. For this reason, cleaning up brownfields so that they can be ready for new development should be a nationally supported (by rural, suburban, and urban constituents) effort. At the same time, there are simply not enough resources to return all brownfields to de facto greenfield status. The marginal costs of site remediation from a lower level suitable for an industrial or commercial use to a higher level suitable for residential use can be extraordinarily high. Therefore, policymaking and remediation decisions must weigh the benefits and costs of cleanup standards for specific sites that take into account their positive and negative externalities as well as market potential.

Understanding the full extent of its brownfield problem is a genuine challenge for any community. Counting only those that make it onto federal or state environmental agency lists will not enable cities to make a reliable inventory. Although a significant part of the brownfields problem is invisible to the public eye, local economic developers need to understand that the private development community has an increasingly sophisticated understanding of both the visible and invisible pattern of brownfields. Furthermore, under present policy and market conditions, this is resulting in an overwhelming preference to take new development projects to the greenfields, which creates even greater challenges for those economic developers who work in the central city and inner-ring suburbs.

Brownfields emerged as a major new challenge for local economic developers only during the decade of the 1990s. This challenge is arguably greatest for local economic developers in central cities and inner-ring suburbs. In this chapter, we seek to provide insight to these local economic developers on the following:

- The issues that brownfields pose for promoting sustainable economic development

- How to characterize the extent of a community's brownfield problem

- The range of barriers that thwart efforts to clean up and redevelop brownfields

- Current gaps in resources and approaches for promoting equitable brownfield redevelopment

- Examples of local economic development efforts of brownfield redevelopment

The success of local economic development efforts in bringing about urban revitalization requires that the planner/practitioner be able to identify known and potential brownfield sites, and to be able to navigate the myriad liability issues and potential environmental issues associated with their redevelopment. Urban brownfields should be cleaned up so that they generate property taxes, make maximum use of existing infrastructure, and provide residents appealing, safe places to live and work. Furthermore, by cleaning up brownfield sites that really are contaminated, local economic developers improve the redevelopment chances for brownfield sites that are not contaminated but, nonetheless, have been adversely affected by the climate of distrust among lenders and developers (Leigh & Hise, 1997). In the end, overall redevelopment potential will be maximized.

Barriers to Brownfield Redevelopment

A typical redevelopment project faces common barriers such as financing and project marketability that are often surmountable. However, when contamination complicates a project, the barriers become considerably more difficult to overcome.

While brownfield issues permeate all aspects of land use and development, they pose especially difficult problems for urban areas because they contain the highest concentrations of brownfield properties. As these brownfield properties are more likely to be idle, abandoned, or otherwise underused, they contribute to the overall blight of urban areas. These sites are also usually the locations of the highest concentrations of poverty and crime, making it even more important that these properties be returned to successful uses to address these other, equally pervasive urban issues. Contaminated land that hinders urban redevelopment efforts as well as regional efforts to manage overall growth and development has become an important policy concern among economic development researchers (Leigh, 1994; Bartsch & Collaton, 1995; Black, 1995a; Bartsch, 1996; Dinsmore, 1996; Davis & Margolis, 1997).

The barriers to brownfield redevelopment can be divided into five key areas:

1. *Liability:* The legal issues surrounding brownfields are complex. Multiple parties (past and present owners) can be found liable. Multiple layers of government can be involved in efforts to assess and resolve liability.

2. *Information:* Information on locations and levels of contamination can be closely guarded and difficult to obtain. Current owners can be reluctant to reveal this sort of information for fear of liability repercussions. In addition, information on effective remediation strategies is minimal, which further limits brownfield redevelopment potential as only the most familiar technologies are considered and they often end up being the most costly.

3. *Cost:* Environmental site assessments and remediation can drive up the cost of a project to the point at which it is no longer financially feasible. It is common for the cost of the cleanup to exceed the actual value of the site. The full costs of cleanup are often incalculable at the outset of a project, thereby contributing to the uncertainty barrier discussed next.

4. *Time:* Contamination hinders redevelopment efforts by lengthening the development period and raising overall development costs to levels that developers can find unacceptable.

5. *Uncertainty:* The unknowns of liability, cost, and time combine to create an uncertainty that can prove to be the most difficult barrier to overcome. Ultimately, uncertainty is what forces investment away from brownfield redevelopment and toward greenfield development.

A successful community brownfield redevelopment program is one that minimizes the five barriers and provides incentives for redevelopment over new development. Features of this sort of policymaking reduce the uncertainty surrounding brownfields by making information and funding readily available while clarifying the liability associated with contaminated properties that, in turn, reduces the time involved in bringing these properties back into active use. While environmental concerns remain central to brownfield issues, the five factors just mentioned reveal that economic interests are also critical considerations. Exactly whose economic interests are served—the classic local economic development question—is also a critical consideration that we address later in the chapter.

Legislation and Programs
Affecting Redevelopment of Brownfields

In this section, we provide an overview of the federal and state legislation and programs that affect the redevelopment of brownfields. Significant public effort has been expended over the last decade seeking to ameliorate the unintended consequences environmental regulations have had in biasing development activity away from brownfields. For those engaged in local economic development, however, it is important to realize that the monies available in these programs represent a tiny fraction of what is needed to clean up and redevelop the nation's brownfields. Further, there is no guarantee that even this small amount of funds (e.g., the $200,000 EPA brownfield pilot grants that went to 50 communities in 2000) will have ongoing authorization. However, even if the present level of public funding for brownfield redevelopment were to continue, it is simply not sufficient to address all of the nation's brownfields.

A recent U.S. Conference of Mayors' (2000) study of the effects of brownfields on 231 cities across the United States found that there were more than 21,000 such properties reported.[1] These properties encompassed more than 81,000 acres and, assuming their full redevelopment potential, were estimated to have a cumulative potential gain in tax revenues of $878 million (U.S. Conference of Mayors, 2000). These contaminated properties affect entire neighborhoods with blighted and vacant lots providing more visible evidence of public and private investment neglect. Thus, an important challenge to local economic developers will be finding other sources of funding for brownfield development.

Furthermore, as we discuss later in this chapter, because there are limits to resources available for brownfield redevelopment, as well as competing sources for all economic development monies, local economic developers and their communities must understand that resolving the nation's and each community's brownfield problem does not mean that all brownfields are fully remediated to greenfield status. Instead, careful planning is required to match the level of cleanup of a site with its potential uses, or what is called risk-based remediation. Furthermore, to do so in a sustainable and just economic development manner, planners and their communities must pay attention to the distributional impacts of their decisions (disproportionately redeveloping high over low-to-no value market sites, and sites in stronger neighborhoods over ghetto neighborhoods). Much of the actual public response to brownfields has focused on the economic efficiency of cleanup and redevelopment of individual properties, as opposed to how remediation can enhance the overall vitality of the neighborhood (Black, 1995b; Iannone, 1996; Simons & Iannone, 1997). As a result, the low-to-no value market brownfields found in the most distressed neighborhoods are proving to be the most difficult properties to develop, for reasons extending beyond the contamination (Leigh, 2000).

Federal Initiatives

The primary federal law affecting brownfield redevelopment is the Comprehensive Environmental Response, Compensation, and Liability Act (CERCLA). The initial intent of CERCLA was to promote cleanup of contaminated properties and to provide opportunities for the EPA to recover cleanup costs from all potentially responsible parties (PRPs). These PRPs can include past and present property owners, lending institutions, and developers even if they did not contribute to the contamination (Congressional Research Service, 1995). Because of fear of being assigned liability as a PRP under CERCLA, interested buyers are wary of taking title, and banks are reluctant to lend on previously developed properties that might have contamination (Leigh, 1994; MacFarlane, Belk, & Clark, 1994; Portney & Probst, 1994).

The EPA is the lead federal agency concerned with brownfields. The EPA is at the center of the nation's environmental policies and programs and is the obvious agency to address the remediation of contaminated property. However, the agency cannot effectively manage all the issues related to the complex process of brownfield redevelopment. While the remediation of a brownfield site requires sound environmental techniques and knowledgeable environmental specialists, a successful redevelopment program involves factors outside of the EPA's area of specialization that require incorporation. To improve the effectiveness of brownfields redevelopment, several other federal agencies are working in cooperation with the EPA to facilitate the reuse of contaminated properties. The Department of Housing and Urban Development (HUD), the Department of Commerce, the Department of Treasury, the Department of Labor, and the Economic Development Administration have joined forces with the EPA to promote the redevelopment of brownfield sites (Davis & Margolis, 1997). The EPA's programs that address brownfields have evolved over the decade of the 1990s to include a broad menu of approaches. Among these are Brownfield Assessment Demonstration Pilots, Brownfield Clean-Up and Revolving Loan Fund Pilots, Job Development and Training, Brownfield Showcase Communities, Brownfield Tax Incentives, Targeted Brownfield Assessments, and the Brownfield Clean-Up Revolving Loan Program.

The EPA is also seeking congressional approval for the Better America Bonds initiative that provide state and local governments with access to zero interest bonds to enable them to clean up brownfields, as well as protect water quality and preserve open space. The agency's FY2001 Better America Bonds proposal includes $10.75 billion in bonding authority over 5 years.[2] These tax credit bonds provide a deep subsidy for communities compared with a traditional tax-exempt bond because the bonds are interest free. The bondholder receives a tax credit from the federal government in lieu of interest paid by the community.[3]

State Brownfield Initiatives

By the late 1980s and early 1990s, many of the nation's states and cities were wrestling with issues of brownfield properties. Environmental and economic development regulations were in many cases creating increased barriers to the redevelopment of areas with known or suspected contamination. This prompted state initiatives to develop plans to overcome the barriers to brownfield redevelopment. Illinois and Wisconsin were among the first states to offer programs to assist in the redevelopment of brownfield sites. Over the next decade, the number of state programs to assist in brownfield redevelopment increased from 1 in 1989, to 46 in 1998 (Bartsch & Anderson, 1998). We include a discussion of state brownfield initiatives in this chapter to help the local economic developer be aware of outside resources available to resolve local brownfield problems.

State brownfield programs are nearly as varied as the states themselves but commonly contain one or a combination of initiatives to limit legal liability, assist in financing assessment and cleanup, provide technical assistance, or all three. Although each state's program is unique, the programs typically incorporate the three areas mentioned next into their efforts to redevelop brownfields.

Legal Liability Relief

Many state programs offer limited legal liability relief for innocent owners or prospective purchasers. Letters of No Further Action, Covenants Not to Sue, or Certificates of Completion are forms of liability relief that can be provided by a state to a landowner after remediation of a site to state-specific standards. Although liability programs are not offered by all states, they are common and useful strategies to stimulate redevelopment of brownfield sites.

A major problem associated with state liability relief programs is that federal legal liability is not automatically forgiven with the conveyance of state liability relief. This has caused severe problems in the past because landowners could still be held liable for environmental contamination by the federal government even after receiving a release of legal liability by a state program. More recently, the federal government has been working with states to create memorandums of agreement (MOAs). A Superfund Memorandum of Agreement (SMOA) or Memorandum of Agreement (MOA) can be negotiated between a state and the EPA, making the state and EPA partners in the redevelopment process. The EPA has stated that one objective of this program is for regional offices to use the negotiation of voluntary cleanup SMOA as an opportunity to define a division of labor between the region and states. This would occur by defining what kinds of sites can fall within the MOA (U.S. EPA, 1996).

Financial Assistance

Financial assistance in the form of loans and grants is an important part of many state brownfield redevelopment programs. Because brownfield sites are typically considered risky investments, it is often difficult to secure funding for redevelopment of contaminated areas.

Low-interest loans can be helpful for redevelopment of sites that will produce sufficient returns to meet remediation costs.

Many of the small and/or low-to-no market value sites that can be found in the inner city or inner-ring suburbs will cost more to clean up and redevelop than they will ever recuperate after resale. Grants will be necessary for most projects in areas in which the demand for property is low and the remediation costs are high. Listed next is a sample of state programs—that provide for areas where these sites are typically found—to show that there is a range of viable funding mechanisms for these areas. However, linking these funding mechanisms specifically to low-to-no market value sites remains a significant challenge to economic developers.

- Connecticut's Urban Sites Remediation Action Program was capitalized with $30.5 million in state bond funds for assessment and remediation of sites in distressed municipalities and targeted investment communities.

- Massachusetts's Brownfield Redevelopment Fund has focused $30 million of funds for low-interest loans and grants for site assessment and cleanup in economically distressed areas.

- Ohio's Urban and Rural Initiative Grant Program provides grants to municipalities or nonprofit organizations in distressed areas. (Bartsch & Anderson, 1998)

Technical Assistance

A variety of state programs provide technical assistance to brownfield redevelopment projects. For example, Wisconsin's Redevelopment Assistance Program offers technical and redevelopment assistance for investigation, cleanup, determination of liability, and other activities related to the redevelopment of a brownfield site.

The state also provides a Business Development Assistance Center (BDAC). This center assists individuals or organizations interested in brownfield redevelopment by coordinating state agency programs and providing this information to the public (Wisconsin Department of Natural Resources, n. d.). The BDAC, and similar programs found in other states, can greatly assist developers in recycling contaminated property by providing in one place the information and requirements of which brownfield developers need to be aware. Determining what regulation and program may be relevant to a specific project can be difficult; this "one-stop shopping" approach can remove some of the uncertainty involved in brownfield redevelopment. These technical assistance programs can encourage the involvement of individuals and organizations that might otherwise lack the knowledge and experience necessary to successfully redevelop contaminated properties.

Characterizing a Community's Brownfield Problem[4]

Hazardous Waste

Many types of commercial and industrial activities make use of hazardous waste or materials that can harm human health and the environment. Hazardous waste is any solid, liquid, or containerized gas that can catch fire easily, is corrosive to skin tissue or metals, is unstable, can explode or release toxic fumes, or has harmful concentrations of one or more toxic materials that can leach out. The EPA defines hazardous waste as toxic, corrosive, ignitable, or reactive materials (Hird, 1994). Hazardous waste can be transmitted through air, water, land, and living animals (Percival, 2000). Examples of hazardous waste include paint removers, varnishes, and motor oil. Some health problems stemming from exposure to hazardous materials include birth defects, infertility, childhood leukemia, heart disease, and respiratory problems (Hird, 1994).

Certain types of historic and present-day land uses are commonly associated with particular kinds of contaminants. For example, gas stations and body shops commonly dispose of auto parts that contain asbestos, like brake pads. Old auto batteries contain sulfuric acid, antimony, arsenic, and lead. If these materials are disposed of improperly, over time they may leak into soil, causing contamination.

Small-Quantity Generators of Hazardous Waste

Large-quantity generators of hazardous waste are facilities producing more than 2,200 pounds each month. These generators are the primary focus of the EPA's Resource Conservation and Recovery Act

(RCRA) program (Hird, 1994). Fortunately, these facilities are not typically found in older communities and neighborhoods. Instead, older communities and neighborhoods have to be concerned with the small-quantity generators of hazardous waste that have not been the main focus of federal environmental policy.

Automotive repair shops, salvage yards, dry cleaners, equipment repair shops, photo shops, construction firms, and Laundromats may look innocuous, but about 1 million tons of hazardous waste comes from these types of businesses. Known as small-quantity generators, these businesses produce between 220 and 2,200 pounds of waste per month. The most common hazardous wastes include those derived from lead batteries, acids, solvents, photographic wastes, and dry cleaning residues (Percival, 2000).

Brownfield sites housing small-quantity generators can create significant problems for local communities. Small sites may contain as much or more contamination as large properties. Prior land uses are an important factor in determining possible environmental contamination of a site. This point is explored further in the following section.

Information Gaps

Lack of information on the extent of their brownfield problem is a barrier for communities seeking to address brownfields in their local economic development efforts. At the national level, contaminant lists exist and are available for the local economic developers who are able to navigate the information network. The EPA has made the CERCLIS (Comprehensive Environmental Response, Compensation, and Liability Information System) database (a database of sites where, under CERCLA, the EPA has conducted some form of site investigation) available on the World Wide Web. In addition, states have maintained lists of leaking underground storage tanks (USTs) that are being monitored. These lists, too, are freely available for those who are familiar with public information sources. Typically, however, information regarding the extent of historical contamination seldom extends beyond these information sources. Furthermore, these sources do not detail all currently known sites with contamination. For example, untold numbers of sites have contamination that is not at a high enough level to make it onto a state or federal list, but which can still pose a barrier for redevelopment. Still other sites do have contamination at levels that could put them on an official list but have not had any property transfer interest necessitating an environmental audit by a prudent potential buyer that might uncover the contamination.

Communities have typically not understood that the extent of their brownfields problem goes beyond CERCLIS or the state UST databases. A significant part of the brownfields problem is, at this point, invisible to the public eye. However, the private development community has an increasingly sophisticated understanding of the pattern of brownfields— visible and invisible. Under present policy and market conditions, this understandably is resulting in an overwhelming preference to take new development projects to the greenfields, further complicating efforts to revitalize central cities and inner-ring suburbs.

In response to the information gap, some communities are establishing brownfields databases to characterize their brownfields situation beyond CERCLIS and UST. Planners in Cuyahoga County, Ohio, in which the city of Cleveland is located, have entered the locations of all known contamination in their community into a Geographic Information System. They then integrated infrastructure and land use into the database to increase their understanding of how contamination is affecting their region (G. Ellsworth, personal interview, July 1997). Planners with the Environmental Protection Division of the state of Texas are eliciting information from owners (on a voluntary basis) of potentially contaminated sites to account for those sites in which contamination has not been officially reported, but is suspected. The city of Pittsburgh created a database, with university support, of available industrial properties on which prospective developers can make customized analyses, including Phase 1 environmental assessments for initial site characterization (Amekudzi et al., 1998). Until such efforts to characterize the brownfield problem become widespread, much will remain unknown about the quantity and location of brownfields in communities of different sizes, states, and regions of the country.

We argue here that communities require a comprehensive understanding of their brownfields problem if they are to succeed in guiding future development that helps to contain urban sprawl, does not exacerbate existing spatial inequities between poorer central-city and inner-ring suburbs and wealthier newer suburbs, and seeks to redress existing inequities that are of concern from an environmental and economic justice perspective. Developing this necessary understanding requires an analysis of current and prior land uses of the community because cities need to be able to recognize their *potential* as well as known brownfield problems.

The potential problem is estimated by identifying properties that have had a previous use that is associated with a high probability of contamination. Understanding prior land uses is difficult, however, because no agency—public or private—produces comprehensive databases on historical land use for every city, or even major city. The primary attempt to associate prior land uses with a probability for contamination was made by Noonan and Vidich (1992). They surveyed environmental engineering firms and established estimates for prior contamination probability based on previous land uses (commercial, industrial, residential, etc.). While their efforts are not exact measurements (thus, there is the potential for survey bias), their probability estimates have been widely cited because no other comparable data have been produced.

Noonan and Vidich's (1992) survey has yielded a comprehensive list of prior land uses associated with possible contamination. For example, it found a 92% probability of finding contamination at oil and petroleum storage facilities, and a 72% probability of finding contamination at sites with prior use as auto repair shops.

The survey results summarized in Appendices 3.1 through 3.6 suggest that the probability of brownfield sites having some form of contamination is high. However, it is important for local economic developers to recognize that the characteristics of former commercial and industrial sites vary greatly. Evaluations of brownfield sites must be made separately, based on information from historical records, Sanborn fire insurance maps, and regulators such as the EPA and state Departments of Environmental Protection.

To inventory and assess brownfield sites for redevelopment potential, local economic developers need to catalog past land uses of properties in their communities. Doing so may require researching the land's use as far back as the late 1800s (Leigh & Coffin, 2000). They must also investigate the regulatory history of the properties and investigate their current uses. Essentially, local economic developers need to conduct preliminary environmental site assessments, also known as ESAs. These are described in the next section on Environmental Site Assessments as Phase 1 ESAs.

Environmental Site Assessments

Investigating and assessing a property for possible environmental contamination can be a complex and time-consuming process. Determining whether contamination exists requires initiation of a three-part environmental assessment on the property in question. This three-part process is called an environmental site assessment (ESA). The ESA is a study generally conducted as part of a real estate transaction—for instance, during the sale or refinancing of a property (Solo, 1995).

1. *Phase 1:* Phase 1 ESAs do not involve site investigation or collecting soil or water samples, but rather, reviewing historical background information on individual properties. The first step in an environmental site assessment is to gather as much information about the site and any activities that have occurred there. The cost of a typical Phase 1 study ranges from $1,000 to $5,000 (U.S. Congress, Office of Technology Assessment, 1995). Qualified engineering firms can be hired to undertake environmental assessments, and lending institutions typically require Phase 1 ESAs from such firms when considering loans on existing commercial and industrial properties. As we noted in the previous section, however, economic development planners can conduct (unofficial) Phase 1 ESAs when planning priorities for economic development revitalization strategies.

2. *Phase 2:* A Phase 2 study can follow Phase 1 work and can include both a review of historical information as well as sample collection and site evaluations. Phase 2 activities may include collecting isolated individual soil and water samples on a site, identifying potential contaminants, and developing a formal work plan for assessing the property. This would include a timetable for investigating the extent of contamination on the site, possible costs associated with the removal and treatment of any contaminants, and a schedule for final completion of the cleanup. The cost of a Phase 2 investigation typically ranges from $50,000 to $70,000.

3. *Phase 3:* Phase 3 entails the actual cleanup of the site. This involves such activities as removing any barrels, drums, or containers of hazardous waste; treating or disposing of any contaminated soil; and removing any contaminants from underground aquifers, streams, or rivers. The cleanup of a site may entail full cleanup, essentially bringing

the site to near greenfield status, or it may entail risk-based cleanup strategies that are tailored to what the site's redeveloped use will be.

As we stated at the outset of this chapter, the marginal costs of site remediation from a lower level suitable for an industrial or commercial use to a higher level suitable for residential use or recreational use can be extraordinarily high. In the case of a former coal gasification site located in Athens, Georgia, the cost of remediation to a level four (industrial use) was estimated to be $2 to $8 million, whereas the cost of remediation to level two (residential use) was $100 million (L. Bradbury, personal interview, February 6, 1997). The reality is that most brownfield cleanup will need to be risk based because there are simply not enough public or private dollars to bring all brownfields back to de facto greenfield status. Phase 3 ESA costs will typically exponentially exceed those of the Phase 2 costs associated with any particular property. Phase 3 costs begin in the tens of thousands of dollars and can go into the hundreds of millions of dollars.

Cleanup of Contaminated Sites

The cleanup of brownfield sites is a complex undertaking, and many physical, chemical, and biological treatments are currently being developed to remediate contaminated sites. Prior to the passage of the Resource Conservation and Recovery Act (RCRA) in 1976, the most common technique for disposing of hazardous waste was landfilling. Landfilling is still used when small quantities of contaminated soil are present on a site and in cases where landfilling is the least expensive option. However, many new remedial technologies have been developed, including chemical, physical, biological, and thermal treatments. Furthermore, from a sustainable development viewpoint, landfilling is the least desirable option for hazardous waste disposal. It displaces the problem to another location, and it does not minimize the burden for future generations of the current generations' waste generation patterns. While it is true that some communities operate landfills for hazardous waste as an economic development strategy, it is, at best, a dubious strategy.

There are three different categories of remediation treatments: (1) excavation and disposal, (2) containment, and (3) treatment. The first category, excavation and disposal, addresses the removal of the hazardous substance and contaminated materials from the site to either an on-site or off-site landfill. However, some hazardous substances can be eliminated through incineration instead of landfilling. Any ash that remains will then be disposed of in an authorized facility.

The second cleanup category, containment, is aimed at preventing existing contamination from becoming worse. Both soil and water can become contaminated through industrial and commercial activity. Plumes of contamination can move through groundwater. Large areas within these underground aquifers, like underground rivers, can become contaminated if substances travel through the soil into the groundwater. Such contamination can travel long distances underground, and it often becomes difficult to trace the contamination of groundwater back to a particular site. Cleanup of groundwater is both costly and time-consuming. It is much more difficult to remediate soil and groundwater contamination, as opposed to dealing with soil contamination only. If testing of soil samples indicates the presence of hazardous chemicals on a property, cleaning up the site so that it can be redeveloped is sometimes as simple as covering the contaminated soil with a layer of dirt. This way, the area can simply be capped off to ensure that the contaminated soil remains on the site.

The third category, treatment, seeks to render contaminants to a state that is no longer hazardous. Many types of treatments are used, including soil washing, bioremediation, and incineration of soil.

Methods of bioremediation are specific to the kinds of contaminants on a site. Bioremediation involves the use of naturally occurring microbes to break down hazardous compounds. "Soil washing" is a technique for cleaning contaminated soils that uses a machine to separate toxic substances from sand ("Soil Washing Proving a Good Bet," 1993). This sand is then washed with a cleaning solution to make sure that any contaminants clinging to the fine particles of sand are washed away.

While the ideal goal may be to return a contaminated site to greenfield condition, the reality is that there are almost no cases where it is possible to do so as the cost of complete cleanup is simply prohibitive. The resulting use and value of the cleaned land will never be large enough to cover the cleanup costs. This is especially true for sites found in the areas of economic decline of our inner cities and inner-ring suburbs. Therefore, each site requires careful consideration of the costs and benefits of remediation in terms of how much cleanup, at what price, and for what use to determine the best level of cleanup, or what we previously identified as risk-based remediation.

Brownlining, Environmental Justice, and Sustainability Issues of Brownfield Redevelopment

Arguments that a process of "brownlining" or "environmental red-lining" was taking place in many urban areas, whereby developers and financial institutions would not consider brownfield redevelopment projects, helped to generate the federal and state brownfield programs that exist today. Parallels were drawn to the original redlining practice of financial institutions and to real estate agents excluding areas of a city from home finance eligibility and home sales promotion because they were "minority" neighborhoods. The potential cost of liability under Superfund and other legislation that extends liability for cleanup of contaminated properties to both "innocent" purchasers and financiers alike has acted as another form of redlining or exclusion. Although environmental redlining may not be explicitly racist, often minority neighborhoods have been excluded from development opportunities because they are located near, or contain within them, urban brownfields.

Efforts to improve both the economic status and quality of life of residents living in areas containing contaminated land has been undermined by this process of brownlining, in which developers and lenders simply will not consider properties with any suspicion of environmental contamination. The safe bet is to develop on greenfield land, and many developers and financiers will not consider anything else. In a poll conducted by the American Bankers' Association, 43% of lenders reported they had stopped making loans to companies whose activities are associated with environmental contamination, like tool and die shops, bottling and canning plants, high-technology metal fabricators, and utilities (Bartsch & Collaton, 1995). In the last half of the 1990s, in particular, the number of federal, state, and local brownfield programs developed to overcome greenfield development bias has grown significantly.

These programs, however, can do little to affect the single most difficult barrier to brownfield redevelopment: market demand. Policymakers have established a continuum of brownfield redevelopment at which at one end of the redevelopment spectrum are "economically viable sites" with potential redevelopment benefits that outweigh the potential costs for contamination. These sites may be afflicted more by the perception of contamination than by actual contamination. At the other extreme lie "economically nonviable" sites with limited redevelopment benefits that clearly fail to cover reclamation costs, regardless of contamination. Although no one explicitly argues that sites at the less viable end of the spectrum should be ignored, their often lower priority in terms of allocation of limited economic development dollars means that they are not being addressed (Iannone, 1996; Davis & Margolis, 1997).

Much of the actual public response to brownfields has focused on the economic efficiency of cleanup and redevelopment of individual properties, as opposed to how remediation can further the overall revitalization of the community (Black, 1995a; Iannone, 1996; Argonne National Laboratories, 1998). Communities typically consider brownfield properties in isolation, as specific problems with specific remedies. Their approach can be characterized as that of "triage," defined here as a market-based approach to brownfield redevelopment in which priority is given to those properties deemed most viable in the marketplace. Argonne National Laboratories (1998), under the guidance of the U.S. Department of Energy, has developed a program for communities that it actually calls "Industrial Triage." The program it recommends instructs communities to evaluate and select sites for redevelopment based on four levels of analysis: (1) a real estate market assessment, (2) an environmental analysis, (3) economic incentives available for remediation and redevelopment, and (4) unique contributions a project could make to the community. Although the intent of the program appears to empower communities, its advocacy of market-driven brownfield redevelopment decisions creates the possibility for continued neglect of the most distressed locations and neighborhoods.

Part of the economic viability that satisfies the public triage approach has to do with the recoverability of cleanup costs, while part has to do with the profitability of the potential reuses of these properties. Both factors are maximized in stronger urban neighborhoods. Thus, it can be argued that the brownfield properties in the most distressed neighborhoods are overlooked for both private and public investment (Leigh, 1994; Iannone, 1996; Simons & Iannone, 1997).

From an economic justice viewpoint, local economic development planners need to ask if economic efficiency is the most appropriate, and the only criterion by which to make public investment decisions for brownfield remediation. Conversely, are there other rationales for directing brownfield public investment in areas suffering the greatest amount of harm from brownfield disinvestment?

To elaborate, the current decision rule for brownfield redevelopment is based on the marketability of the property once clean. Thus, the marketable brownfield properties, while seen as expensive investments, have value almost regardless of the existence of contamination. As economic development planners, we need to consider whether the economic efficiency rationale that underlies the current brownfields policy focus best serves the goals of local economic development. Further, we need to ask whether the "cherry picking" of premium brownfields can remedy the negative and unintended consequences of past industrial practices that have been the focus of environmental justice concerns. Should there, instead, be a more comprehensive approach to brownfield redevelopment that focuses policy efforts on the entire community, guided by decision rules based on the economic effects of brownfield redevelopment for the entire community rather than just the individual property? We can call this the community development approach to brownfield development.

The suspicion of contamination has historically had a negative effect on land values, on both the source property and those properties surrounding it. For example, it has been found that the stigma associated with land contamination can reduce property values by up to 94% (Patchin, 1988, 1991, 1994). The level of market impact depended on the severity of the contamination identified on or near the subject property (Patchin, 1988, 1991, 1994). The community development approach to brownfield redevelopment can be seen as a potentially more equitable way to approach the issue that also results in a more economically efficient way to address brownfields. The current triage approach measures the effects of redevelopment in isolation, considering individual properties rather than entire neighborhoods. The measurable positive externalities, however, extend well beyond property being redeveloped. These positive spin-off effects include such things as a clean and appealing environment, which can attract new economic activities on and around these properties and stimulate further positive economic activity in the brownfield's surrounding neighborhood. In the next section, we profile local programs that represent examples of community development approaches.

Local Brownfield Programs[5]

Many municipalities and counties have developed brownfield remedi-ation and redevelopment programs. Although often funded by state and federal programs, these programs have been modified by local govern-ments to fit the specific needs of their brownfield problems. Many municipalities started their brownfield initiatives by focusing on publicly owned or tax-delinquent properties. The advantage of this specific focus is that a municipality has easy access to the site, as well as control over decision making.

Kalamazoo, Michigan

Kalamazoo's brownfield program evolved over the decade of the 1990s. It first established a Brownfield Redevelopment Initiative (BRI) in 1994 to redevelop publicly owned land with three major goals:

- To retain control of land development and ensure its fit with the master plan and community's idea of what they wanted their town to become

- To clean up and sell publicly owned property and sell those prop-erties to increase the local tax base

- To use state money to clean up property (state program requires the property be publicly owned for money)

Kalamazoo had a collection of land it had acquired over the years through sheriff sales of tax-delinquent properties. The city has the right to purchase the property, at a very low price, before the properties are offered for sale to the general public. Kalamazoo has used state money for its public property project and is now looking at bringing private properties into the project to help private property owners (City of Kalamazoo, n. d.).

Kalamazoo received an EPA Brownfield Pilot grant in 1996. With this grant, they developed an inventory of publicly owned brownfield sites. The city has identified, prioritized, and acquired brownfield sites that have reverted to public ownership because of the failure of previous owners to pay property taxes. City staff have assembled resources to conduct site preparation activities such as demolition and environmental site assessment. Through public input sessions with community stakeholders, the city has sought to ensure that high-quality purchase and development agreements are negotiated between it and perspective developers (City of Kalamazoo, n. d.).

The following is a list of benefits of the Brownfield Redevelopment Initiative identified by the City of Kalamazoo (n. d.).

- Community benefits
  1. Protection of public health and a cleaner environment
  2. Tax-base enhancement by finding productive uses for neglected sites
  3. Job creation and retention
  4. Spin-off redevelopment and stronger neighborhoods
  5. Creation of an alternative to urban sprawl and the loss of open space
- Benefits to developers who successfully redevelop these sites are significant
  1. Reimbursement for eligible environmental expenses (baseline environmental assessment, "due diligence," and additional response activities)
  2. State Single Business Tax Credit for up to 10% of investment in property improvement ($1 million upper limit on credit)
  3. Resources for enhancing private investment with public improvements (City of Kalamazoo, n. d.)

In 1997, Kalamazoo expanded its brownfield program with the creation of Brownfield Redevelopment Districts. Brownfield authorities were established with a board of directors to oversee these districts. These authorities formulate a redevelopment plan for a project that is then sent to the state for approval. The authorities then acquire properties through purchase with funds from the Kalamazoo general fund for economic development, or through donations of public or private properties. The city's Brownfield Redevelopment Authority and District allows the city to capture taxes on designated brownfield sites to cover the costs of testing and cleaning up environmental problems. Developers who invest in these sites are given significant tax credits (City of Kalamazoo, n. d.).

MacKenzie Bakery is an example of a completed BRI project. The bakery was constructed on 1 acre of a 3.5-acre brownfield site. Site preparation and land acquisition for the entire brownfield cost the city $89,000. The 1-acre parcel was sold for $25,000. Instead of the sale proceeds going back to the city's budget for other projects, $1,250 has been set aside to create a job-training program and the remaining $23,750 is being used for public infrastructure improvements around the site (B. Gordon, personal interview, October 1998).

Community input has been a key feature in the city's brownfield redevelopment process. Because the projects did not have preidentified end users, the city used stakeholder groups (community members, environmental specialist, city staff, etc.) to design projects and plan for redevelopment. In essence, the Brownfield Redevelopment Districts emphasize community development over traditional economic development goals. That is, the choice of end product does not focus solely on the marketability or economic benefits of a particular piece of property (B. Gordon, personal interview, October 1998).

Emeryville, California

Although the city of Emeryville has successfully redeveloped some of its large brownfield properties, small parcels and the remaining large parcels had proved more difficult to redevelop. This was due to time and regulatory uncertainty as well as the issue that the estimated returns were insufficient to offset transaction and potential cleanup costs. Redevelopment investment was difficult to obtain with such risks and uncertainties. In 1996, Emeryville applied for and received a Brownfield Redevelopment Pilot Grant for expediting the redevelopment of the more difficult-to-redevelop brownfields through a Groundwater Management Program. The city decided to identify pathways of contaminated groundwater to humans and the environment with the goal of solving problems regionally and cooperatively, rather than on a piecemeal, site-by-site basis. Cleanup levels could be established for the entire city, based on proposed land use.

The Groundwater Management Program is designed to protect public health, deep groundwater resources, and the ecological resources of

San Francisco Bay, while providing regulatory relief and more cost certainty for property owners, developers, and responsible parties. Recognizing that brownfield redevelopment is often a time-consuming process that must be tackled in steps, the city has developed and implemented portions of a multistep program that allows the city to achieve this goal. Highlighted next are the steps Emeryville undertook to facilitate the redevelopment of brownfield sites (City of Emeryville, 1998).

Develop Financial Incentives. Emeryville investigated numerous financial mechanisms for assessment and remediation of brownfield properties. Brownfield redevelopment expenses, such as soft costs (i.e., assessment, engineering), remediation, and infrastructure, can be funded through many sources, each with its own conditions. These come in the form of loans or grants. Using several of these sources, Emeryville and its partners redeveloped several significantly contaminated large brownfield sites. They used tax increment financing, assessment districts, an EPA Brownfield grant, tax incentives, and responsible parties to help assess and remediate the City's brownfields.

Emeryville also interviewed property owners and developers to determine if a loan-grant program for site assessment would stimulate redevelopment of *small* sites. It learned that although the programs would be welcome, what most stakeholders need first is a simplification of the regulatory process and city assumption of groundwater management. Furthermore, site assessment financial assistance would not be useful unless it is backed up by remediation financial assistance.

Manage the Regulatory Process. Emeryville has provided many property owners and developers with an understanding of the regulatory process and assistance in obtaining sign-off. In addition to regulatory assistance, it may be necessary for cities to proactively intervene in the confusing regulatory process. Emeryville seeks to simplify the property redevelopment process by combining an environmental sign-off process within the regular planning approval process. Emeryville is assuming authority for processing environmental sign-off for soil and groundwater investigation and cleanup projects within city limits. The city will step in as an "intermediary" between owners-developers and regulatory agencies. Regulatory agencies will retain approval authority, but the sign-off will be processed by the city. This step reinforces the one-stop shopping concept and, it is hoped, will facilitate redevelopment of smaller properties.

Establish the City as a "Center Under Memorandum of Understanding." Under this step, owners-developers are to submit soil and groundwater reports to the city. The city manages the regulatory process and seeks approval from the appropriate regulatory agency. Under the agreement, the city will be empowered to "sign off" on sites that do not require extensive regulatory oversight, and it will serve as a liaison for more complicated cases.

The Louisville-Jefferson County Landbank Authority

Land bank authorities are generally established by local governments (city or county) to address urban blight and promote redevelopment. They acquire tax-delinquent properties with the goal of returning them to productive use. They are typically nonprofit entities empowered by state or local governments to waive or forgive back taxes owed on a property.

To date, very few localities have used land banks to promote brownfield redevelopment although their potential to do so is great. As the following discussion of the Louisville Jefferson County Landbank illustrates, land banks could receive vacant brownfield properties that are tax delinquent and environmentally contaminated; arrange for remediation and sale of such properties thereby putting them back into productive use; and take a proactive stance in acquiring vacant known or potential brownfields that are tax delinquent.

The Louisville-Jefferson County Landbank Authority in Louisville, Kentucky, appears to be unique as a land bank authority that has specifically created a mechanism to address brownfields (Leigh, 2000). The Landbank Authority is a cooperative partnership made up of four taxing districts: the City of Louisville, Jefferson County, the Commonwealth of Kentucky, and the Jefferson County Board of Education. Each of the four taxing districts provides a representative to make up the land bank authority's governing board (F. Nett, personal interview, October 1998, May 1999).

Established in 1989, the Landbank Authority was not originally created to address brownfield properties. In 1997, however, the Landbank

Authority accepted a resolution that permitted any of its member organizations to acquire brownfield properties through the land bank's property acquisition channels. Stipulations were made in the resolution that the risks associated with the acquisition of a brownfield were "confined to the member acquiring the property through the Landbank Authority program; and . . . that this member will indemnify the other members of the Landbank Authority" (Louisville/Jefferson County Landbank Authority, 1997). This resolution also established four criteria that must be met before a brownfield property can be acquired through the Landbank Authority:

1. The property is abandoned.
2. The property is known to be contaminated by hazardous material or substances.
3. The costs of remediating the contamination on the property are considered to be reasonable in relation to the value of the property.
4. There exists a potential purchaser for the property after the remediation of the contamination (F. Nett, personal interview, October 1998, May 1999).

In 1994, Louisville created a citywide brownfield redevelopment initiative as part of its application process for Empowerment Zone status from HUD. Though not successful in the national Empowerment Zone competition, the city was named an Enterprise Community (EC) and awarded $3 million. In 1995, Louisville received a Brownfield Pilot Site grant. With the $200,000 from this grant, the city launched site assessment activities in targeted areas in the EC. One of those sites was the Ni-Chro Plating facility. City officials believed that because of the interest shown by a local business owner, the site would make a good pilot project (Pepper, 1997).

The Ni-Chro Plating site was a chemical plating facility that had been abandoned for several years. The U.S. EPA performed an emergency contamination removal at the site in 1987 at a cost of $168,000. The site had been abandoned for nearly 7 years when, in 1994, a neighboring business owner, the Louisville Dryer Company, approached the city with an interest in purchasing the property. The company stipulated its purchase of the property was contingent on the city resolving the property's environmental issues and providing future liability protection after the property transaction. The company employed approximately 25 people at the time of its first meeting with the city and through the expansion of the facility, hoped to hire at least 10 to 15 more employees. The city decided to assist the owner with his expansion project, rather than forcing the Louisville Dryer Company to relocate to another area (Pepper, 1997).

Along with the environmental concerns of the prospective purchaser, financial problems were also a major barrier to the reuse of the site. The EPA had assessed almost $200,000 in liens for its past cleanup activities to the property, although the assessed value of the property was only $35,000. Because of the discrepancy between costs and property value, the site would never be attractive to a prospective purchaser unless the burden of the liens could be removed. After researching the liens, the EPA determined that the statute of limitations on the liens had expired, so they were forgiven. At this point, the Landbank Authority acquired the property through tax foreclosure with knowledge of the contamination. Additional assessment and sampling of the site was conducted by the city after the title was transferred to the Landbank Authority. The environmental assessment process revealed nominal contamination, and the city proposed a risk-based remediation strategy. It suggested leaving low-level contaminated soils on-site that could be capped by gravel and the building to be constructed on the property. The owner of the Louisville Dryer Company worked with the community to resolve some design concerns and reduce the impact of the facility on the area (Pepper, 1997).

Funds for the Ni-Chro project were assembled from several sources. The Louisville Brownfield program received strong support from the EPA in the form of its Brownfield Pilot grant and an additional $75,000 "seed grant" that was established through the state. The city has secured $6.5 million in preliminary private and foundation commitments to purchase stock in a new Community Development Bank. While the Community Development Bank funds are not specifically directed toward brownfield projects, the city hopes that the more investor-friendly atmosphere developed through the bank's incentives to stimulate the redevelopment of the area will encourage brownfield cleanup in the EC area. In 1995, Louisville received $7 million from HUD in support of the EC plans. A portion of these funds will be used for brownfield projects (F. Nett, personal interview, May 1999).

Through the use of the Louisville/Jefferson County Landbank Authority to clear the title for the Ni-Chro site, a highly improbable redevelopment effort succeeded. Due to the large debt owed on the property, along with the environmental contamination and liability issues, it is highly unlikely this site could have been redeveloped without the public sector's provision of strong technical and financial support.

Expanding Land Banks' Roles in Brownfields: Strategies for Funding Acquisition, Remediation, and Redevelopment

The staff and program of the Louisville/Jefferson County Landbank Authority, Inc., are funded entirely out of the city of Louisville's General Fund. An alternative method of funding land banks has been established in Massachusetts. Both the Nantucket and Martha's Vineyard Land Banks are funded by 2% transfer fees that provide funds to pay off bonds that were issued for the purchase of open space. The Nantucket Land Bank was established in 1983 by a special act of the Massachusetts Legislature. The Martha's Vineyard Land Bank was created one year later.

The Massachusetts State Legislation authorized the land banks to issue bonds and notes, including notes in anticipation of bonds, for the purpose of acquiring land and interest in land for conservation. This 2% transfer fee is levied on all but a few exempt property transfer types and must be paid to the land bank before the deed can be registered. If the purchaser of a property refuses to pay the transfer fee to the land bank, a lien in favor of the land bank is issued on all property and right-to-property belonging to the purchaser (personal communication, Krauss, 1999). The mission of both land banks is to preserve open space, rather than to redevelop brownfield properties. However, local economic developers may want to explore the use of a transfer fee by land banks to create a fund for brownfield remediation in their communities.

Such a fee in combination with the Brownfield Redevelopment Fund described later could provide valuable tools for local economic developers to use in redeveloping the low-to-no market value brownfield sites that are currently being overlooked in the dominant triage approach described earlier in this chapter. Again, the example comes from the State of Massachusetts. In 1998, the Massachusetts Government Land Bank merged with the Massachusetts Industrial Finance Agency to form the Massachusetts Development Finance Agency (MDFA). In the same year, the Brownfield Redevelopment fund was established and placed under the management of the MDFA. Thirty million dollars were appropriated by the Massachusetts state legislature for the Brownfield Redevelopment Fund during the commonwealth's 1998 regular session. The fund is to be kept segregated from other agency funds. Proceeds and income from the fund are to be invested and reinvested—through the provision of loans and grants to finance ESAs and environmental cleanup actions—to promote development in economically distressed areas of the commonwealth (Mass. Ann. Laws, ch. 29, sec. 2W, 1999).

Lessons for Economic Development Practice

The discussion earlier in this chapter on the pervasiveness of contamination suggests the inevitability of local economic developers in central cities and inner-ring suburbs finding that their communities have brownfield problems. Resolving these problems is an important component of an overall community revitalization strategy.

We argue here that to promote effective and equitable revitalization strategies, local economic developers need to understand the pattern and extent of their community's brownfield problem. To do so requires that they inventory and map their known and potential brownfields. If only brownfields in the most marketable sections of the community are cleaned up, redevelopment efforts may act as a force for widening inequalities between areas within the community. Consequently, local economic developers need to be aware of the community's needs for brownfield redevelopment in areas that are overlooked in the current brownfield policy climate.

Brownfield redevelopment districts, such as Kalamazoo's, are a promising concept for maximizing the number of brownfields that are cleaned up and the overall revitalization impacts they can have for a community. Individual brownfield sites are usually not found in isolation. They are more likely to be found in areas that contain other brownfields. Thus, the brownfield redevelopment district approach helps to reveal the "forest for the trees" when coordinating public efforts and funds for remediation and redevelopment.

In creating a brownfield redevelopment strategy, local economic developers should evaluate where the private sector will take the initiative to redevelop brownfields and use public programs and incentives to encourage their initiative. In doing so, local economic developers might want to consider adopting the city of Emeryville's approach of "share the risks, share the rewards." That is, Emeryville believes the private sector, in exchange for the community's acceptance of bearing the residual risk for risk-based cleanup for its brownfield redevelopment project, should return a portion of its savings on remediation expenses to the community. Requiring the private sector to do so could also help fund other brownfield projects of the city.

From Emeryville's initial experiences with brownfield redevelopment, we highlight two other recommendations that can enhance local economic developers' understanding of what will be needed in creating their own communities' brownfield strategies. First, because brownfields often coexist with other problems, such as a lack of infrastructure and community services, localities should remove these deficiencies and offer incentives for non-brownfield-related development costs. Second, in general, small sites and projects will need proportionately more loans, grants, technical assistance, or all three than larger sites and projects.

That smaller sites may need greater assistance suggests they will tend to be found less marketable than larger sites. Using a cost efficiency perspective alone might bias economic development away from smaller sites. However, an understanding of the community's pattern of brownfields may illuminate value in expending the proportionately greater resources required on these sites to promote greater equity and environmental justice.

As we stated at the outset of this chapter, a successful community brownfield redevelopment program is one that minimizes five barriers: liability, information gaps, costs, time, and uncertainty. Thus, the successful program must reduce the uncertainty surrounding brownfield redevelopment by making information and funding readily available, and by clarifying the liability associated with contaminated properties that can reduce the time required to carry out the projects. The final lesson we offer to economic developers engaged in brownfield redevelopment is to consider setting up one-stop business centers (which can be located at the state level as in Wisconsin or at the local level as in Emeryville and Chicago) to package efforts aimed at minimizing the five barriers. Having a visible center staffed by those who can walk the private developer (small or large), or community development group, through the myriad complexities of brownfield redevelopment could go a long ways toward encouraging more brownfield redevelopment projects and ensuring that they are completed as quickly as possible.

Appendix 3.1	Probability of Environmental Contamination, by Industrial Land Use Categories
Industrial Land Use Category	*Probability of Contamination (in percentages)*
Former coal gas plants	99
Fuel distributors	99
Chemical distributors	99
Incinerators	99
Plastic manufacturing	95
Refining	95
Metal plating	90
Chemical manufacturing	90
Metal finishing/tool and die	90
Heavy industrial manufacturing	88
Paper manufacturing	88
Tannery	87
Circuit board manufacturing	85
Metalworking and fabrication	83
Machine Shops	80
Electronics assembly facility	80
Agricultural mizers/formulators	80
High-technology manufacturing	80
Electronics manufacturing	79
Industrial parks	75
Automotive assembly facility	75
Light industrial manufacturing	75
Textile printing and finishing	65
Fabric dyeing establishments	50

SOURCE: Derived from Noonan and Vidich (1992).

Appendix 3.2	Probability of Environmental Contamination, by Transportation Facilities, Utilities, and Storage Facilities Land Use Category
Transportation Facilities, Utilities, and Storage Facilities Land Use Category	*Probability of Contamination (in percentages)*
Electric utility	95
Hazardous waste storage/transfer	95
Oil and other petroleum storage	92
Landfills	90
Power plants	88
Tank farms	85
Waste treatment plants	85
Railroad yards and rights-of-way	82

(Continued)

Appendix 3.2 Continued

Transportation Facilities, Utilities, and Storage Facilities Land Use Category	Probability of Contamination (in percentages)
Vehicle maintenance facility	82
Refuse recycling facility	80
Trucking terminal	65
Resource recovery facility	60
Highways	40
Warehouses	35
Gas utilities	35

SOURCE: Derived from Noonan and Vidich (1992).

Appendix 3.3   Probability of Environmental Contamination, by Commercial Land Use Category

Commercial Land Use Category	Probability of Contamination (in percentages)
Auto salvage yards	95
Gas stations	88
Furniture repair and stripping	85
Junkyards	79
Dry cleaning	74
Auto repair	72
Electrical/plumbing/HVAC service	60
Auto dealership	53
Photographic	50
Retail property	25

SOURCE: Derived from Noonan and Vidich (1992).

Appendix 3.4   Probability of Environmental Contamination, by Location of Vacant Land

Vacant Land Category	Probability of Contamination (in percentages)
Urban vacant/abandoned land	85
Rural vacant property	20

SOURCE: Derived from Noonan and Vidich (1992).

Appendix 3.5	Probability of Environmental Contamination, by Residential Properties and Offices	
Residential and Office Land Use Category	*Probability of Contamination (in percentages)*
Residential property	20
Office (nonmanufacturing)	13
SOURCE: Derived from Noonan and Vidich (1992).	

Appendix 3.6	Probability of Environmental Contamination for Hospitals, Laboratories, and Research Facilities	
Hospital/Lab/Research Land Use Category	*Probability of Contamination (in percentages)*
Chemical research facility	70
Pharmaceutical establishments	50
Research facilities	40
Hospitals	20
SOURCE: Derived from Noonan and Vidich (1992).	

Notes

1. This study follows the initial study on brownfields generated by the U.S. Conference of Mayors in 1996. The 1996 study only surveyed 39 cities, and the results revealed similar but less comprehensive results.

2. The proposal requires a change in the tax code; new tax legislation must be enacted before the program can go forward. Better America Bonds proposals were introduced in both houses of Congress in 1999 and are still under consideration—the Better America Bonds Act (H.R. 2446) and the Community Open Space Bonds Act (S. 1558). Because there was no major tax bill in 1999, no action was taken on this legislation (J. Maas, personal communication, March 29, 2000.)

3. For example, the issuer of a million-dollar bond would save more than $900,000 over 15 years by issuing a Better America Bond instead of a tax-exempt bond, assuming annual payments into a sinking fund at 6.5% (J. Maas, personal communication, March 29, 2000).

4. This section draws from Leigh and Hise (1997).

5. This section draws from Leigh (2000).

4

Industrial Retention:
Multiple Strategies for
Keeping Manufacturing Strong

Industrial retention includes a range of activities designed to respond to the needs of local businesses, including business visitation programs, infrastructure improvements, safety concerns, technical assistance in modernization, and employment training. What is important about these efforts is that they put a city in a more proactive position in shaping its economic base, rather than responding to the threat of plant relocations or closings. Industrial retention has the potential to build toward each of the three macro principles identified in Chapter 1.

Manufacturing jobs pay better than most service sector jobs available to people with a high school education or less. The availability of manufacturing jobs, then, can reduce income polarization in a city or region. To the extent that manufacturing retention reuses existing facilities that would be abandoned and prevents greenfield development and the wasteful duplication of physical capital associated with it, it represents a sustainable development strategy.

Why Manufacturing Matters

Plant closings and massive corporate downsizing from the mid-1980s to the mid-1990s had experts, politicians, and the average U.S. citizen convinced that the nation's manufacturing era had truly ended. As headlines announced major plant closings and massive job cuts and as employment vulnerability spread from blue- to white-collar workers, the postindustrial service economy seemed to firmly grab hold. From our vantage point at the turn of the century, it is clear the death knell for U.S. industry was too hasty. Manufacturing still accounts for 15.8% of the nation's employment and is still an important component of many urban economies (Table 4.1).

Manufacturing jobs are important to a local economy because they pay higher wages overall than service industries for people with comparable skills (Table 4.2). Further, as an export sector, manufacturing creates employment in other sectors.[1] In 2001, the average weekly earning for manufacturing workers was $595.03 compared with $441.94 for service workers. The difference in wages between these two sectors has remained constant since 1990 (U.S. Department of Labor, Bureau of Labor Statistics, 2000).[2]

The location of manufacturing employment has shifted dramatically. Since the 1980s, industrial displacement has been widespread in many central cities (Betancur & McCormick, 1985). Beginning in the 1940s, the level of suburban manufacturing growth exceeded that of cities (Stanback & Knight, 1976). By the 1960s, both suburban and urban economies were restructuring, with continued growth of high value-added manufacturing in the suburbs, and decline of manufacturing in cities, where business services were expanding (Stanback & Knight, 1976). By the late 1980s, the location quotient, a measure of economic concentration, for manufacturing in many suburbs was higher than their corresponding central cities (Stanback, 1991). This trend continues today.

Over the past 30 years, many cities have been trying to slow the relocation of manufacturing to the suburbs. As manufacturing moves to even more distant sites, inner-ring suburbs will have to focus more on industrial retention as well. Retaining some firms, such as those moving to underdeveloped countries for cheap labor, is beyond the scope of any local economic development strategy. Yet there are a significant number of manufacturers for which central cities and inner suburbs are still a desirable location. Retention efforts can be focused on these firms.

The importance of industrial retention programs is perhaps understated in economic development practice today. Despite considerable

TABLE 4.1   Manufacturing Employment in Selected U.S. Cities
(by percentages)

	1990	1992	1994	1996	1998	2000
Los Angeles	20.2	18.8	17.3	17.0	16.8	15.4
Chicago	18.1	17.2	17.0	16.5	15.9	14.9
Boston	14.4	13.8	12.7	11.9	11.4	10.5
Detroit	23.0	22.1	22.4	21.8	21.0	20.8
Dallas	16.4	15.3	14.8	14.3	13.7	12.7
Houston	11.0	10.9	10.7	11.0	11.0	10.1
Atlanta	12.6	12.3	11.9	11.3	10.9	10.1
Seattle	19.8	18.6	16.9	16.1	16.9	14.1
Cleveland	22.7	21.4	20.6	20.1	19.4	18.7
Minneapolis/St. Paul	19.1	18.5	17.9	17.4	16.8	15.9
National Averages	17.4	16.7	16.0	15.5	14.9	14.0

SOURCE: Adapted from U.S. Department of Labor, Bureau of Labor Statistics
(1990-2000a).

TABLE 4.2   Average Weekly Earnings of U.S. Production Workers,
1996 to 2001 (in dollars)

Industry	1996	1998	2000	2001[a]
Construction	602.94	646.13	701.90	693.12
Manufacturing	531.65	562.53	596.77	585.68
Transportation	571.82	604.75	624.47	644.23
Wholesale Trade	492.92	538.88	584.43	610.27
Retail Trade	230.11	253.46	273.11	282.35
Fire	459.16	512.15	547.04	584.87
Services	382.00	418.58	453.88	476.44

SOURCE: Adapted from U.S. Department of Labor, Bureau of Labor Statistics. (1990-
2000b).
a. Preliminary data.

evidence that business attraction inducements only redistribute jobs
rather than create new development, they remain the dominant
approach to economic development for a number of reasons. Job creation is the mantra of economic development. Elected officials are under
pressure to be aggressive in attracting new employers. Although successes may be few, new business locations make headlines and create
good publicity for elected officials. Doing deals is the glamorous side of
economic development compared with the nuts-and-bolts relationship

building and land use planning of industrial retention. Although deal making is eventually part of industrial retention, incentives to firms are focused around a broader strategy rather than being offered on a company-by-company basis. Attraction efforts are wasted if a supportive environment for manufacturing is not maintained.

The beginning point of most industrial retention strategies is assessing the needs of a city's or region's diverse manufacturing firms. Manufacturing used to be dominated by a few large employers. Now any given city or suburb with a manufacturing base is more likely to have many small- and medium-sized businesses. Typically, more than one program is needed to respond to differing needs for infrastructure and site improvements, neighborhood crime prevention, workforce training, and expansion. However, the planning process is more complicated than identifying needs and responding to them, as illustrated in the cases in this chapter.

We focus on industrial retention efforts in Chicago, and more briefly on Seattle, Washington; Portland, Oregon; and Euclid, Ohio, an inner-ring suburb of Cleveland. We emphasize Chicago because it has, through its six programs (Table 4.3), one of the most comprehensive industrial retention strategies in the country. Collectively, the programs illustrate an ongoing commitment to manufacturing retention spanning more than several mayoral administrations. We emphasize Chicago's planned manufacturing districts (PMDs) because this strategy is gaining popularity as land use competition intensifies in many cities and the approach is being replicated. PMDs are "super" manufacturing zones that provide insurance to manufacturers so their investments are being protected.

The protracted battles that developed over the designation of each of Chicago's four PMDs illustrate the conflicts that can emerge when different types of development compete for the same space. When such competition occurs, the city finds itself in the position of choosing between two seemingly incompatible goals—maintaining or attracting middle-class residents and maintaining businesses that provide jobs for working-class residents. The story of how the PMDs were implemented reveals this type of planning is as much a political process as it is a land use planning exercise. The planners and community organizations involved in implementing PMDs saw them as one component of a broader strategy for maintaining jobs that provided living-wage employment for Chicago residents with a high school education or less.

TABLE 4.3   Chicago Manufacturing Retention Programs

Industrial Retention Program	Objectives
Local Industrial Retention Initiative (LIRI)	City's Department of Planning and Development designates community organizations to be the liaison between the City and firms in their area.
Planned Manufacturing Districts (PMDs)	City delineates areas with strict manufacturing zoning that prevent encroachment of competing uses, primarily residential.
Industrial Corridors	City established 22 industrial corridors and focuses on their infrastructure and physical needs. LIRI organizations are present in many of the corridors.
Tax Increment Financing Districts (TIFs)	City designates blighted area to attract investment by returning tax revenue generated through development above the original assessed value for improvements in the area.
Brownfields Initiative	City initiative to address physical, legal and other barriers to assembling and improving brownfield property for development.
Industrial Parks	City is developing four parks by updating infrastructure and services in them and offering development subsidies.

SOURCE: Compiled by author.

Chicago's comprehensive industrial retention strategy did not emerge as a totality; rather, it evolved over time. Each program was the brainchild of a new mayor or commissioner. As each new program was implemented, it was not necessarily connected to previous programs. It was only after several years of implementation that the separate programs were linked to form a comprehensive strategy. The continuity of planning staff over several administrations was one factor that allowed integration of separate programs into a coherent strategy. Successful implementation of Chicago's industrial retention strategies could not have been possible without a broad network of community-based organizations that assumed most of the day-to-day planning work. These community organizations still work as agents of the city through a program called the Local Industrial Retention Initiative. Because it is unlikely the PMDs would have happened without this program, we begin with a brief description of it. The PMDs have been strengthened and supplemented with other industrial retention programs, which are discussed after the story of how the PMDs were developed.

Industrial Retention Strategies in Chicago

Manufacturing employment in the Chicago standard metropolitan statistical area (SMSA) has been growing since 1990. From 1990 to 1996, 1,542 firms began operations in the metropolitan area, all locating outside the City of Chicago. In fact, the City experienced a loss of 455 manufacturing firms during this time period. Manufacturing employment in Chicago fell by more than 50,000 jobs from 1990 to 1996 (Illinois Department of Employment Security, 1996). The surrounding metropolitan area saw an increase of approximately 32,500 jobs. Thus, the manufacturing jobs located outside the City increased from two thirds in 1990 to three fourths by 1996 (Table 4.4).

Clearly, action was needed to protect the jobs that did exist because they provided many job openings through replacement jobs even though the sector was not growing. There was widespread recognition that the City needed to stem the outflow of manufacturers.

The Local Industrial Retention Initiative

An essential task of industrial retention is knowing the needs of various manufacturers. Keeping "a finger on the pulse" of manufacturers requires more staff and resources than most planning departments have available. Because a city the size of Chicago has 3,800 manufacturing firms employing over 170,000 people (Illinois Department of Employment Security, 1996), one can see that understanding the manufacturing base is no small task.

The City created the Local Industrial Retention Initiative (LIRI) to address this problem during the Harold Washington administration in 1984 as one component of a broader economic development strategy to balance the development needs of neighborhoods and downtown (see Clavel & Wiewel, 1991; Mier, 1993). Previously, the Department of Economic Development had focused on being alert to early warnings of plant closings and attempts to force companies that did not keep promises to repay the City for incentives.[3] LIRI was a move from a more combative approach that was not producing results to one based more on public-private partnerships with a decidedly community-based orientation.

TABLE 4.4   Chicago SMSA Manufacturing Firms and Employment,
1990 to 2000

	Chicago City	Remaining SMSA	Total SMSA
Firms			
1990	4,420	9,052	13,472
1996	3,965	11,049	15,014
2000	3,443	11,106	14,459
Percentage of change	−22.1	22.6	7.3
Employment			
1990	216,190	447,087	663,277
1996	166,139	479,633	645,772
2000	147,092	483,166	630,258
Percentage of change	−31.9	8.1	−5.0

SOURCE: Adapted from Illinois Department of Employment Security (1996, 2000).

The LIRI program provides funds to established community organizations to be the liaison between the City and businesses in their geographic area (Mier & Moe, 1991). LIRI groups provide a range of services including planning, needs assessments, maintaining inventories of existing facilities and, through frequent company visits, acting as an early warning system for the City. As business needs are identified, LIRI groups present recommendations to the appropriate city departments for action. The program allows the City to keep abreast of, and to respond to, industry needs at a relatively low cost. LIRI groups have several sources of funding for their work. Funding from the City ranges from $55,000 to $85,000 per year. Currently, 12 organizations are funded in different geographic areas of the City.

As one of the first LIRI groups designated in 1984, the Greater North Pulaski Development Corporation illustrates the type of planning tasks the groups accomplish. Greater North has made 4,200 personal contacts with industry representatives, intervened 3,850 times to provide assistance, made 350 local site location referrals, provided 980 employment referrals, given financial or technical assistance to 560 firms, and identified and contacted 700 new companies since it started. Executive Director James Lemonides (personal interview, May 1998) explains that interventions range from such mundane items as speeding up the removal of abandoned automobiles to group organizing efforts to remove blight from the community, or helping a company expand within the community when leaving was seen as the only alternative. One of the most valuable contributions the LIRI groups make, according to Lemonides, is to offer a vehicle through which businesses can participate proactively in the development of their communities.

The difference between the LIRI program and traditional business visitation programs is that the latter are often operated by citywide planning departments or extension services and concentrate solely on the needs of business. The LIRI groups place business needs in the context of other local businesses, the broader community, and the employment needs of nearby residents. The LIRI groups were instrumental in developing the planned manufacturing districts.

Planned Manufacturing Districts

A planned manufacturing district (PMD) is designed to prevent competing land uses, specifically residential and commercial, from encroaching on manufacturing areas. The PMD replaces zoning on land already zoned for manufacturing by creating an industrial area in which land use is specifically defined and uncontestable. Zoning codes in many cities are contestable on a plot-by-plot, or "spot," zoning basis. In contrast, as defined by Chicago statute, zoning within a PMD can only be changed with a majority vote by City Council. In enacting enabling legislation for a PMD, a city makes the statement that the public benefits from manufacturing uses supercede the public benefits of competing uses.

Expansion of residential development in largely manufacturing areas creates several problems. First, residential invasion almost guarantees increases in property taxes for manufacturers. Developers buy property at manufacturing rates, convert it, and sell it as residential. As property values are reassessed for tax purposes, the land is valued at what it could bring on the residential market—often a difference of up to $40.00 per square foot. Land values are evaluated every 3 years, based on current value according to possible land uses and the current market value of surrounding properties. Thus, as more properties are sold at the higher residential rate, the assessed value of property remaining in the original use rises as well.

Second, if manufacturers want to purchase adjacent land for expansion, it is likely that they would be outbid by residential developers (King, 1988). Because a residential use gains more immediate profit per square foot of land than an industrial use, the residential land developer is willing to pay more to realize a higher rate of return.

Third, residential zoning conversions within a manufacturing zone force businesses to make operational changes, often at considerable expense. Even if operations changes are not required, businesses may not be able to operate as efficiently when residential uses are permitted nearby. Manufacturers use land in ways that frequently conflict with residential ideals. Companies often operate on 24-hour schedules, loading and unloading trucks that are noisy at night and block traffic during the day. They use sidewalks for loading and unloading, conflicting with residents' desires for street parking and a quiet neighborhood. Parked cars lining a street hinder the movement of truck traffic. As parking pressures increase, some residents park in areas designated as loading and unloading zones. Some manufacturing processes produce noxious odors and create clutter from crates, boxes, and waste materials. Residents in these areas can file nuisance complaints if they find the conditions intolerable. As in most states, Illinois state zoning code requires manufacturers within 300 feet of residential property to adjust their operations to accommodate residential needs, even if the manufacturing use preceded the residential. These changes can be very costly to the business as they might require constructing walls or fences or setting back equipment and processes from the property boundaries. In many cases, it may be cheaper for a company to sell the property and rebuild elsewhere rather than make the necessary changes and continue to deal with hostile neighbors. For many businesses, this may mean a move from the city to less developed suburban areas.

The First Battle: Creation of the Clybourn PMD

Planners in the City's Department of Economic Development and community-based organizations began documenting development pressures that were displacing manufacturers in the early 1980s (Giloth & Betancur, 1988). Their studies found that small manufacturers were more likely to relocate to suburban locations than in the city (Ducharme, Giloth, & McCormick, 1986). The City's policies on manufacturing retention were contradictory. Although a set of programs focused on infrastructure improvements and financial incentives to maintain and attract manufacturers, the City facilitated residential and commercial development that competed with manufacturing by offering zoning changes and other financial incentives (Giloth & Betancur, 1988). PMDs would be the strategy the City would take to protect manufacturing more consistently.

The need to create Chicago's first PMD emerged in the mid-1980s when real estate developers began requesting zoning changes to convert industrial and warehouse buildings into residential lofts. Some planners in the Department of Economic Development unsuccessfully attempted to fight the zoning changes that allowed loft conversion.[4] When development conflicts emerged in the area north of the Loop known as the North Branch Industrial Corridor (see Figure 4.1), the right combination of community organization and support from the City emerged to develop a coherent response.

The North Branch Industrial Corridor consisted of three concentrations of manufacturing: Clybourn Avenue, Elston Avenue, and Goose Island. Adjacent to the area is Cabrini Green, one of the City's largest public housing projects; Lincoln Park, one of the wealthiest neighborhoods; and West Town, a Latino and Eastern European working-class neighborhood. The corridor provided approximately 30,000 well-paying manufacturing jobs. In 1988, manufacturing wages in the corridor ranged from $19,600 to $26,200, depending on the industry, while nonmanufacturing wages in Chicago averaged $10,700 (Chicago Association of Neighborhood Development Organizations, 1988; City of Chicago, 1988). Although industrial zoning districts had been in place for many years, the City had no focused strategy for responding to pressures facing businesses located within them.

The first industrial-to-residential conversion occurred in 1983, with the development of Clybourn Lofts, an abandoned factory converted into 57 loft condominiums. Despite signing purchase agreements with statements of awareness about being in an industrial district, Clybourn Lofts residents began complaining about their manufacturing neighbors. After Clybourn Lofts was completed, several more residential conversions were granted. Well-connected developers found it easy to appeal to their alderman for spot-zoning changes that allowed them to purchase property zoned for manufacturing and then convert it to residential use. By 1988, several more zoning changes were granted. Manufacturers in

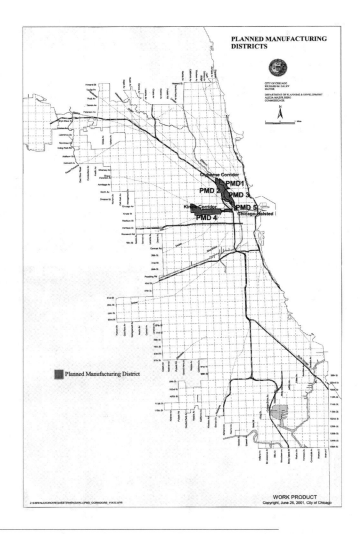

Figure 4.1.   Chicago's Planned Manufacturing Districts
SOURCE: Copyright by the City of Chicago; reprinted by permission of the City of
          Chicago Department of Law.

the area became concerned about nuisance complaints and increased
taxes. In the mid-to-late 1980s, industrial land sold for $6.00 to $9.00 per
square foot and retail property for $12.00 to $20.00. Residential property commanded as much as $40.00 per square foot (Ducharme, 1991).

In Chicago, property value increases are especially significant as manufacturing property is taxed at 36% of value, while residential property is taxed at 16%.[5] Because Chicago has a head tax[6] in addition to a higher commercial property tax rate than the surrounding counties, many businesses were understandably concerned they would be taxed out of business if their property values rose.

Soon after the Clybourn Lofts conversion, the area's LIRI group, the Local Employment and Economic Development (LEED) Council (the economic development arm of the local YMCA), began organizing the corridor's businesses, residents, and labor unions to ask the City to develop a coordinated land use plan for the area that would protect the interests of manufacturers. A research and organizing agenda was coordinated by the Research and Development Division of the City's Department of Economic Development, the LEED Council, and the Center for Urban Economic Development at the University of Illinois at Chicago. Donna Ducharme, founder and then director of the LEED Council, relates that the organizing effort took 3 years: "One for the idea to gain credibility, one to build support in the proposed districts, and one to forge the coalition needed to get the PMD approved" (personal interview, May 1998).

Residents in the surrounding neighborhoods had to be convinced the PMD would not create nuisance effects. The key to gaining their support was creating a buffer zone for commercial development along Clybourn Avenue that would prevent commercial development from creeping into the manufacturing and residential areas on either side. Unions voiced their support as the job retention and expansion potential became evident.

The LEED Council data convinced Alderman Martin Oberman that an industrial market existed for the Clybourn properties being converted to residential uses. In 1985, he ordered a moratorium on zoning changes and asked the City and the LEED Council to develop a land use plan for the area. The plan for creating a PMD in the Clybourn Corridor was completed in July of 1986. The PMD could have a manufacturing core and a buffer that would allow commercial development in addition to manufacturing. Approval of the PMD-enabling legislation, however, was delayed because City officials realized the problem of industrial displacement extended far beyond the Clybourn Corridor. The LEED Council was instructed to draft enabling legislation for multiple districts. Once this legislation was passed, the Clybourn PMD ordinance was signed by Mayor Eugene Sawyer in October 1988.

The LEED Council documented continued industrial displacement and pushed to create PMDs in two additional manufacturing strips, Goose Island and Elston Avenue, which were established in 1989. Ironically, the legislation was signed by Mayor Richard M. Daley, who campaigned against PMDs. Since then, Daley has come to appreciate the importance of keeping manufacturing in the City and is still a forceful advocate of PMDs.

Although three battles to create PMDs were won, as demand for residential development in Chicago continued to expand, the war moved to other fronts as illustrated in the recent designation of the City's fourth PMD, the Kinzie Industrial Corridor.

The Kinzie Corridor PMD

The Kinzie Industrial Corridor, spanning 613 acres on Chicago's west side, is a long-standing manufacturing area. The corridor is home to 430 industrial firms that provide over 13,700 jobs with an estimated payroll of $480 million. They contribute approximately $685,000 in employee "head" taxes to the City and pay $6.8 million in property taxes. According to unpublished documents of the Industrial Council of Northwest Chicago, the corridor's land is zoned 62% manufacturing, 29% nonindustrial, and 9% vacant. The Kinzie PMD contains a variety of firms: small light manufacturing, heavy manufacturing, and the Fulton-Randolph Market, a six-block wholesale produce, fish, and meat market.

It can be argued that the creation of the Clybourn PMD created a demand for residential development in other neighborhoods close to the Loop, including the Kinzie Corridor. City planners estimate 10,000 units of housing could have been built in the Clybourn Corridor if the PMD had not been enacted. In the early 1990s, developers began converting some of the abandoned warehouses and manufacturing buildings in the Near West Side into loft apartments and condominiums. Urban "pioneers" began moving in. Trendy restaurants began opening in the heart of the Fulton-Randolph Market area. Manufacturers and wholesalers began to feel threatened.

In 1995, the ward's alderman, Walter Burnett, and the Industrial Council of Northwest Chicago (a membership organization of the corridor businesses and an LIRI group) noticed a rise in requests for zoning changes. By this time, Ducharme had been appointed as deputy director of the Department of Planning and Development Industrial Services Division. She allocated funds to complete a strategic plan for the Kinzie Industrial Corridor to document whether there was a need for another PMD. The plan, based on a survey of members and focus groups, was completed in October 1996. It identified five initiatives for infrastructure and physical improvements and proposed creating a PMD.

It took more than 2 years from when the process was initiated to the signing of the ordinance to create the Kinzie PMD in March 1998. Dennis Vicchiarelli, then director of Land Acquisition and Management for the City's Department of Planning and Development, was charged with facilitating the process. Vicchiarelli (personal interview, May 1998) estimated there were at least 25 versions of the PMD map as specific properties were added or deleted. Different constituencies coalesced to support or oppose the PMD. Over 25 public hearings were held.

The positions of developers, residents, and businesses mirrored those in the Clybourn Corridor hearings. Land speculators who had purchased abandoned properties for residential conversion argued that the derelict multistory buildings were not usable for manufacturing and claimed that if a PMD were imposed, they would not be able to resell the buildings.

Residents were divided. Those in the surrounding working-class neighborhoods favored the PMD because it would maintain jobs. West Town United was created to give a voice to community residents concerned about maintaining the area's ethnic and class diversity and preventing gentrification and job loss.[7] The organization has 20 institutional members, including churches, settlement houses, economic development organizations, and business associations. In contrast, residents in the newly developed lofts and townhouses opposed the PMD, arguing that banning residential development would turn their neighborhood into a lonely residential island in an industrial environment.

Business owners were also divided. Most manufacturers favored the PMD, although a few saw the demand for residential conversion as an opportunity to make large profits on their properties. Aldermen were pressured by developers to oppose the PMD, and for the first time, those who Mayor Daley could count on for support began expressing doubts.

The mayor made it clear the PMD was important to him and personally met with his opponents to ask for their support. The Kinzie PMD was approved by City Council in March 1998.

Industrial Corridors

It is perhaps because of the controversies created around the PMDs that the Daley administration began another industrial retention program rooted in more city-led bread-and-butter land use planning. In the early 1990s, the Department of Planning and Development (DPD) created three regional plans for industrial opportunities covering the north, west, and south sides. The plans detailed how land use planning and zoning could prevent industrial displacement. The overall plan designated 22 industrial corridors throughout the city. The industrial corridors are contiguous areas in which industrial uses are concentrated; an expectation that such uses will continue exists, and little conversion to other uses has taken place. Some of the factors planners considered in defining the industrial corridors were parcel configurations, separation from non-industrial uses, local circulation patterns, and transportation access.

Then Deputy Commissioner Ducharme wanted to expand the corridors beyond an exclusively physical focus. She designated 12 of the 22 corridors as model industrial corridors that would use the LIRI organizations to develop plans for the infrastructure and other needs of businesses. Each model corridor was awarded a one-time $50,000 grant to develop strategic plans for transportation, infrastructure, and human capital development. The planning funds originally came with a commitment from DPD of up to $1.5 million per corridor to seed development projects.

A new commissioner of the department was appointed before implementation began, and the commitment was rescinded. The new commissioner's focus was to create tax increment financing (TIF) districts to finance infrastructure and other improvements in manufacturing areas. Some of the corridors were able to complete projects including site beautification, improved signage, and land purchases for truck staging. As the emphasis shifted to TIFs, most of the priority items in the corridor plans were incorporated into the priority list for TIF funding.

Tax Increment Financing

Chicago is using TIFs to spur manufacturing development, both in conjunction with PMDs and industrial parks. TIFs, an economic development financing tool used to attract private investment to blighted areas, gained popularity in the 1980s as federal funds for urban economic development declined. To establish a TIF, local governments create an authority to administer the redevelopment of designated blighted areas. Property is taxed within the TIF at its assessed value before development. The difference between the original and postdevelopment values generates TIF revenue and becomes the captured assessed value that can be used for improvements such as land acquisition and preparation, road and sewer construction, and streetscaping (lighting, landscaping, parking) (Klemanski, 1989).

With over 110 TIF districts (15 are industrial), Chicago is unrivaled in its use of this economic development tool, a practice that is not without controversy. *Crain's Chicago Business* (Hinz, 2000) reports the share of the City's property tax base in existing and pending TIFs almost tripled between 1995 and 1997, to $2.05 billion, or 6.7% of Chicago's total 1995 equalized assessed valuation of $30.38 billion. The numbers suggest Chicago is more heavily invested in TIFs than any other major U.S. city. The extensive use of TIFs was originally controversial because tax revenues could be diverted for up to 22 years from general revenues, which fund schools and other public services. In cities like Chicago with daunting public education challenges, this diversion is particularly problematic. Many of the TIFs are now being structured to protect some or all of the tax revenues for education. Further, $100 million in unspent TIF proceeds have been advanced to the Chicago Public Schools for new school construction.

When used in conjunction with PMDs or industrial parks, the TIFs are proving to be powerful development tools. In February of 1998, the Chicago City Council approved three new industrial TIF districts and later voted to overlay a TIF on the West Pullman Industrial Park. Because of the decline of federal funding for urban development, TIFs may be one of few remaining tools for financing needed improvements in industrial areas. Chicago needs these improvements to expand manufacturing. The *Illinois Real Estate Journal* (Fingeret, 1998) reports, "More than 82 percent of the 1,018 buildings in the districts are

35 years of age or older; 17 percent of the buildings and 201 acres of land are vacant; and almost 882 buildings are deteriorating" (p. 14). Further, the property appreciation in the districts is not consistent with that of the rest of the city. The City estimates the three new TIFs could create 5,750 to 8,000 new jobs and increase the earned assessed value of their property from $135 million to $255 to $315 million by 2015 (see Arthur Andersen LLP, 1998a).

Chicago, unlike its suburbs, has been hesitant to float bonds to front fund TIFs.[8] Bonds provide the borrowed money needed against the expected increase in tax revenue. Chicago relies more on the "pay-as-you-go" method in which the developer pays for project costs and is reimbursed by the City over time as the tax revenues are realized. Weber (1999) elaborates on the funding gap this creates: "Because developers require larger up-front sums of money than the increments trickling in, they often turn to banks and other financial intermediaries to fill the initial financing gaps" (p. 3). Chicago has addressed this financing gap by using HUD's Section 108 Loan Guarantee funds. Historically, this program has been used as supplemental funding for economic development projects too large for HUD Community Development Block Grant funds. Chicago has worked out a deal with HUD that provides up-front funding for site cleanup of industrial TIFs that keeps payments low in the early years and increases them as the TIF generates more revenue (DeVries, 1998).

Goose Island, which has both a PMD and a TIF, provides an example of how industrial retention programs can complement each other. Because of these programs, more than 1.2 million square feet of industrial space has been retrofitted, developed, or planned on Goose Island, resulting in several new expansions (Arthur Andersen LLP, 1998a). Federal Express recently built a 120,000-square-foot distribution center. Although Federal Express owned the property for 10 years, it did not want to build on it because of needed environmental remediation. The TIF district provided $2 million for remediation and job training, enough of a commitment to spur Federal Express to build. The company added 100 employees, reaching a total workforce of 330, and anticipates 100 new employees will be added yearly for the life of the TIF (Fingeret, 1998). In addition, Republic Windows and Doors is building a 375,000-square-foot office and factory facility, and Sara Lee Corporation consolidated three facilities into an existing 264,000-square-foot building on Goose Island. By combining the programs, the City was able to respond to the concerns of existing manufacturers, create an environment for developing new facilities, and provide a funding mechanism to make it happen.

Of course, the City draws on many sources of funding: federal and state Economic Development Administration (EDA) grants, block grants, and so on. Chicago has been especially adept at combining these sources of development finance. What is most interesting about the TIFs is they are strategic tools that can complement other economic development programs if used judiciously.

Industrial Parks

Even with the PMDs, the demand for industrial space in Chicago exceeds the supply. PMDs were implemented to solve a particular problem, namely, protecting manufacturing from encroaching residential and commercial development. This approach was especially important to manufacturers such as A. Finkl and Sons Specialty Steel, in the Clybourn Corridor. Finkl has been in its location since 1865 and has invested considerable capital in its multibuilding facility. It would not have been financially feasible to relocate its plant and equipment. Finkl worked hard to encourage city officials to support this solution to protecting its substantial capital investment in Chicago.

Other solutions are needed for keeping firms that have more locational flexibility. For manufacturers considering relocating to, or expanding in, the City, the problem is finding developable land. For some firms considering expansion, the feasibility of moving to other parts of the City may be an option, if suitable sites were available. A 1998 report by Arthur Andersen calculated an annual demand of 1.8 million square feet of industrial space through the year 2005 (Arthur Andersen LLP, 1998a), 67% of which is for replacement, expansion, or both of existing firms. Indeed, exit interviews with manufacturing firms leaving Chicago indicate the lack of adequate space for expansion or modernization is a leading reason for their departure—21 of the 24 firms interviewed indicated they moved because they could not expand at their existing site (Boston Consulting Group, 1998). All 21 responded they would have stayed in Chicago had land been available. Approximately 3,866 new jobs could be created, and property taxes of $219.8 million realized, between 1998 and 2005, if the total demand for land were met (Arthur Andersen LLP, 1998a).

The issue may not be land per se, but assembly of available land. The report suggests that to meet ongoing industrial demand, the City must continue to aggressively assemble, zone, clean, and market industrial sites. Further, the report suggests these efforts should focus on reuse of the 4.5 million square feet of industrial space available in the west and south sides of the City rather than protecting space in the already congested north side.

That movement is beginning to occur already as the City develops four new industrial parks. One success is Stiffel Lamp Company, which is one of several firms moving to the near south side's Stock Yards Industrial Park. Before moving, Stiffel sold its six-story building on the near north side to a developer who is converting it to loft condominiums. The City is developing two additional business parks on the west and south sides, including the Roosevelt and California Business Park. The plan is to develop 685,000 square feet of space on the 55-acre site, creating between 1,400 and 1,750 new jobs. Recently, much of the wholesale produce market in the Kinzie Corridor moved to this park. The West Pullman Park on the south side has 160 developable acres of land, with the potential for 1.3 million square feet of new developed space. At capacity, the park would create 3,800 to 4,100 new jobs and retain 500 jobs. In its plans for these multipurpose parks, Arthur Andersen (1998a, 1998b) recommends the City create a central marketing and management center with associated social and business services and include security and job training to connect local residents to jobs. Currently, there are offers on three sites in the West Pullman Park for industrial development.

These plans represent a second generation of industrial parks. The industrial parks that proliferated in small towns and rural areas in the 1970s and 1980s were based on the assumption of "build it and they will come." Few of these parks were successful. The difference in the urban parks being developed in Chicago is that they are in industrial areas where there is documented demand for industrial space. The parks will be marketed to growing sectors such as intermodal distribution, warehousing, and business services. To make the parks attractive, the City is assuming responsibility for security, landscaping maintenance, signage, lighting, and service requests of tenants. By including the marketing, management, and training amenities, the City is positioning itself well for continued expansion of high value-added manufacturing growth.

As residential development continues in and around the Loop, more industrial development is occurring in the industrial parks and other industrial areas of the city. Over 900 acres of abandoned manufacturing land in the Lake Calumet area on the far south side are being redeveloped into an intermodal transportation center[9] and a supplier park to Chicago's Ford assembly plant (Hinz, 2000). The City is particularly well positioned because the railroad freight industry is growing and Chicago has six rail lines (DeVries & Lenz, 1999). The park will consolidate Ford's supplier system. The City and state have put together a $115 million incentive package to create 1000 new jobs (plus 2000 "saved" jobs at the Ford assembly plant) through the two developments. Of the direct expenditures, approximately $43.5 million is for street and roadway improvements, $23 million for workforce development, $18 million for site development, and $2.4 million in energy efficiency improvements to the Ford plant. Indirect savings to Ford are in the form of property tax reductions, tax credits, and energy savings. The energy improvements financed by the city and state will save Ford almost $1.5 million per year. Construction starts in 2002.

The location of the supplier park is particularly rewarding for Chicago because it redevelops an area of the City that has been abandoned for years. The site and the package to develop it for Ford was in stiff competition with a greenfield site near Atlanta's Hartsfield Airport. Without the subsidies, Chicago could not have competed with the greenfield site.

Experiences of Other Cities and Suburbs in Protecting Manufacturing

Chicago was not the first city to use the protected manufacturing district. As part of its comprehensive growth management plan, Portland, Oregon, developed a land use designation in 1981 similar to PMDs—the industrial sanctuary. The state mandated that all cities develop these plans to contain urban sprawl. These homogeneous industrial zones allow commercial or retail development only when they are ancillary to the industries in the sanctuary. Retail buildings in the sanctuaries are limited to 16,000 square feet. The City provides infrastructure improvements needed by manufacturers such as street improvements and waste elimination programs. Of the 16 industrial sanctuaries created in 1981, three have been legislatively removed to accommodate other types of development.

More recently, developers wanting to build big-box retail stores or do residential loft conversions have questioned the benefit of the sanctuaries. Their opposition has not gotten far because converting a sanctuary to another use requires a change in the City's comprehensive land use plan, which must be approved by the state legislature, thus establishing a system of checks and balances. Unlike Chicago's PMDs, a developer can apply for an exception if the property is located on the sanctuary's boundary. The cost, however, is prohibitively expensive. Further, Portland does not have a retail sales tax, so the City is not motivated to support retail over manufacturing uses. The state of Oregon views the industrial sanctuaries as a success and now requires all cities to create some form of industrial sanctuary as a part of their comprehensive land use plans.

In 1994, Seattle, Washington, created manufacturing-industrial centers (MICs), also as part of comprehensive growth management plans mandated by the state to prevent urban sprawl. The Seattle Office of Management and Planning studied various industrial retention strategies and chose to create a hybrid of Chicago's PMDs and Portland's industrial sanctuaries (City of Seattle, 1997). The City's comprehensive plan defines MICs as clearly defined geographic areas with transitional land buffering the borders. Almost 90% of industrially zoned land is now contained within the City's two MICs. The Duwamish and Ballard Inner-Bay North End MICs are approximately 5,000 and 940 acres, respectively. The 1994 comprehensive plan specifies a job creation target of 14,660 jobs in the MICs over a 20-year period.

MICs differ from PMDs in three respects. First, they concentrate manufacturing more than the PMDs. Second, retail uses that function ancillary to industry are permitted, although strict regulations limit the size of retail structures. Commercial offices are permitted, resulting in competition for vacant plots of land. The ambiguous references to ancillary uses in the MIC ordinance make it difficult to exclude these non-industrial uses. Third, they allow conversion of obsolete manufacturing warehouses and buildings for residential use, which is limited to artist live-work studios. New residential construction is not permitted.

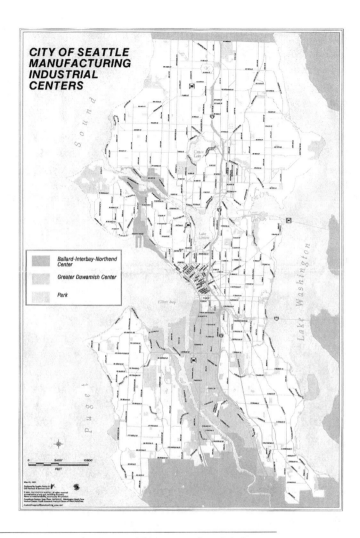

Figure 4.2.   Seattle's Manufacturing Industrial Centers

Residents who move into the live-work buildings must sign a contract acknowledging they are aware of the industrial surroundings and, in any legal dispute, industry has priority within the area. It is not clear whether these agreements are legally enforceable, and the issue has not been tested. Because of these differences, creating the MICs did not produce protracted battles as in Chicago.

As commercial and residential development pressures increase, demand for more restrictive zoning is emerging. In response, the Seattle Office of Economic Development is considering establishing size limitations for retail and office developments within the MICs and using buffers as in the Chicago PMDs to accommodate commercial and retail uses. In both Chicago and Seattle, planned manufacturing strategies have had to be adapted. In Chicago, this has meant adding more programs to support the PMDs and to include other sections of the city. In Seattle, where industrial uses are more concentrated, the exact uses permitted within the districts are periodically evaluated.

Industrial Retention in the Suburbs

Industrial retention is not on the radar screen of many suburban planning departments. Its need will become evident if the trend toward increasing dispersion continues. The industrial parks of many inner-ring suburbs have been in place for 30 to 40 years. Because these landlocked suburbs have little land available for expansion, revitalizing older industrial parks is essential if manufacturing is to remain viable. Elk Grove Village, a suburb of Chicago, has started a revitalization program for its industrial park. It is funded through a telecommunications tax that will raise approximately $2 million annually.[10] Almost one half of the funds are used to upgrade transportation access by widening roads, reconfiguring intersections, creating more parking, and creating turnoff spaces for buses that run through the park. Other needs are for signage and landscaping to compete with the attractiveness of newer industrial parks. A loan fund is available for companies to improve docking facilities and to raise roofs to create high-demand 18-foot ceilings. Further, Village planners are working with local banks to create a cooperative program to increase the low-interest loan fund.

Most Chicago suburbs are using TIFs for commercial, rather than industrial, development. Suburban Bartlett, for example, has created a TIF in an old manufacturing area to construct a retail and office center. Residents are highly supportive because the new development is attractive and the few manufacturing jobs left are not considered vital to the local economy. Indeed, each community must do its own assessment of the need to maintain or expand its manufacturing base. Evidence suggests many inner-ring suburbs are not concerned about manufacturing loss, especially if commercial and office development are viable options.

Most of the suburban manufacturing efforts we uncovered offer the traditional package of location incentives. Euclid, Ohio, a blue-collar suburb of 52,717 people on Cleveland's east side, launched its Euclid Means Business Campaign in 2000. It is a publicity effort using print and radio advertising and promotion brochures to promote redevelopment of a declining industrial corridor on Euclid Avenue. The area is being redeveloped with funds from the U.S. Department of Justice's Weed and Seed program. James Sonnhalter, of Euclid's Department of Community Services and Economic Development, explains it is designed to "weed out bad elements and seed new development" (personal interview, May 2001). A community-policing program has been instituted, and an abandoned apartment complex has been demolished to develop an office and light industrial park. Empowerment Zone inducements are available to firms locating in the area. To date, one company purchased a 4-acre site and broke ground in 2001 to build a 42,000-square-foot facility. The machining company was located in another section of Euclid but had no room for expansion. It is nearly doubling the size of its facility and will expand from 17 to 30 employees. A bigger challenge will be finding a buyer for an abandoned 600,000-square-foot PMX plant. Although Euclid gets weekly inquiries, there have been relatively few takers.

Lessons for Economic Development Practitioners

Chicago has been trying to defy the downward trend in manufacturing employment since the early 1980s. It is difficult to determine the effect of Chicago's long-term commitment to industrial retention. Manufacturing loss is still occurring in the city and growing in the rest of the primary metropolitan statistical area (PMSA) (Table 4.4), but the city's loss may have occurred at a more rapid rate without the programs. Factors favoring the suburban and exurban parts of the metropolitan area are stronger than Chicago's programs. Yet the new industrial parks and industrial TIFs are reinforcing PMDs and encouraging manufacturing expansion in formerly abandoned areas of the south and west sides. Jon DeVries, principal at Arthur Andersen, is convinced the numbers will start climbing in 2002 as the industrial parks fill (personal interview, July 2001).

Through its six industrial retention programs, Chicago is responding to the infrastructure, land use, and physical needs and concerns of manufacturers. The LIRI program is helping to retain small and medium-sized manufacturers whose workforces live nearby and provides a cost-effective way of identifying and responding to the needs of these firms. The PMDs address the specific problem of residential and commercial invasion into industrial areas. The industrial parks are providing needed developable space in abandoned but viable areas of the City. They target existing firms that need to expand but are attracting new development as well. The industrial TIFs are providing a needed financing tool in an era of limited federal funding for urban development. Further, the Brownfield Initiative augments industrial retention efforts. Model industrial corridors is the only program that has not been fully integrated into the broader industrial retention mix. Instead, it has been supplanted by a focus on industrial parks funded with TIFs. The multipronged approach suggests no single program is sufficient to address an economic development need as important as industrial retention.

Chicago's comprehensive industrial retention strategy did not emerge as a totality, but rather, it evolved over time. Each program was the brainchild of a new mayor or commissioner. The emphasis on manufacturing started in the Washington administration continued in the Daley administration. Despite initial hesitancy, Daley quickly caught on to the importance of the LIRI program and PMDs and continued to support them. He and the commissioners he appointed also wanted to develop their own programs, and thus industrial parks, TIFs, and the Brownfield Initiative were added to the industrial retention program mix. Practicing planners know all too well that highly effective programs can disappear with the coming and going of different administrations and department heads. In Chicago, industrial retention has been a consistent focus over three administrations and five commissioners.

As each new program was implemented, it was not necessarily connected to previous programs. It was only after several years of implementation that the separate programs were linked to form a comprehensive strategy. The continuity of planning staff over several administrations was one factor that allowed integration of separate programs into a coherent strategy. The long-term involvement of Chicago's broad network of community organizations through the LIRI program has provided a continuity of commitment to manufacturing. Further, the involvement of community organizations reveals economic development is not confined to planning staff.

In fact, the involvement of Chicago's well-established network of community-based organizations (CBOs) ensured that industrial retention would serve the interests of working-class people and their neighborhoods, as opposed to focusing exclusively on corporate subsidies. It was CBOs (in conjunction with planning staff in the Washington administration, many of whom had come from CBOs) that argued that the frequent granting of spot-zoning variances by aldermen represented continued favoring of the interests of real estate developers over neighborhood development. The ability to grant zoning variances gave aldermen considerable power. PMDs, which curb this privilege, represent a shift of power in zoning from the council (elected officials) into the hands of planners and the community.

The expanded use of TIFs is creating questions over reduced funding for other public services. The LIRI groups and other CBOs have raised public awareness of the trade-offs associated with this economic development tool. These critics are not against the use of TIFs under any circumstances and, in fact, are supporting projects financed by the industrial TIFs. However, the expansive use of this financing mechanism in areas that are not blighted or experiencing disinvestment is being scrutinized closely. In this sense, CBOs provide a system of checks and balances to ensure economic development planning serves the public interest. More TIF deals are now being structured to protect tax proceeds for schools.

Most decisions over land use are inherently political. PMDs prioritize the needs of manufacturers over other constituencies. During the Washington administration, the issue was the relative importance of supporting manufacturers in the neighborhoods over big-ticket downtown development projects. Although Daley originally campaigned against PMDs, the presence of a strong political constituency for the program influenced him to change his mind. As the conflict over PMDs continued in the Daley administration, the presence of competition between two of Mayor Daley's key concerns—bringing the middle class back to the City and maintaining and expanding manufacturing—has become evident. These goals do not have to be in competition. In Seattle and San Francisco, restrictive leases seem to provide enough protection against residents filing complaints against manufacturers. In Chicago, urban planner Dennis Vicchiarelli (personal interview, May 1998) points out that judges tend to rule conservatively on issues of land regulation, meaning they are more likely to rule in favor of citizens in land use disputes. By focusing more on the industrial parks, the City has been able to direct future development and reduce these conflicts.

Jon DeVries (personal interview, July 2001), a principal at Arthur Andersen, notes that the focus on industrial parks represents a maturing of Chicago's brownfield redevelopment efforts from remediating single sites to focusing resources on opportunities that allow significant land assemblies that can compete with greenfield sites. Development of the industrial parks, however, has increased the use of development subsidies dramatically, as illustrated in the Ford supplier park. The subsidies have more restrictions than those typically associated with such developments. For example, the Ford agreement has clawback provisions that require the company to create a minimum of 500 full-time jobs by the end of 2006 and to maintain these jobs through 2011. Further, Ford must create a minimum of one million square feet of building space. If these commitments are not met, Ford must pay back financing proportionate to the employment shortfall and repay the City for infrastructure and road improvements. Ford has also guaranteed that all new jobs within the supplier park will be covered by United Auto Worker (UAW) union contracts and to maintain the existing 2,800 union jobs at the main plant.

The political environment that motivated the creation of protected areas for manufacturing in Portland and Seattle was quite different than in Chicago. The sanctuaries and MICs were created to comply with state mandates for comprehensive planning to manage growth and contain sprawl, goals motivated by an environmental consciousness that does not exist to the same degree in many states. The mandate to create manufacturing districts was reinforced by a public consensus on both environmental and economic development policy. It appears, however, that this tool by itself is not enough to stem the commercial and residential invasion that started in the mid-1990s.

An important lesson learned from the story of PMD implementation is planners must be cautious in adapting "best practice" programs from other cities. In some cases, replicability is more direct. Pittsburgh and Cleveland replicated the LIRI program in almost the exact form as in Chicago. Seattle planners visited Chicago but adapted the PMD concept considerably in forming MICs. Chicago's experience is relevant to other cities. The lessons, however, are not in lists of "best practice" programs, budgets, or zoning maps. Rather, they are in uncovering the key actors and political conflicts, the historical relationships among different organizations and agencies charged with implementation, and opportunities and strategies for intervention based on existing state and local programs and policies. What cities can learn from each other is how to organize for industrial retention (or any economic development strategy) rather than the current version of any particular program. The particular mix of these factors is what shapes how industrial retention unfolds in other places.

Equity and sustainability are at the core of industrial retention efforts. Chicago's comprehensive industrial retention strategy fares well under our definition of economic development. To the extent that well-paying manufacturing jobs are retained and expanded, more residents will have opportunities for earning a living-wage. This was the explicit goal of the CBOs and Washington administration planners who used the LIRI program and the PMDs to achieve this end. It is not clear, unfortunately, that the resources invested in retention have had significant impact. The Brownfield Initiative and the industrial park development strategies have an explicit focus on environmental sustainability, but to the extent that all industrial retention programs encourage in-place over greenfield development, they promote sustainable development. This goal is explicit in the MICs of Seattle and the industrial sanctuaries of Portland. In Chicago, it is less so, although the City's brownfield site was chosen over a greenfield site in suburban Atlanta.

The examples discussed in this chapter demonstrate the importance of land use planning as a tool of economic development generally, and to industrial retention in particular. In a global economy, however, availability of land is only one factor in industrial location. For most cities and suburbs, industrial retention needs to be part of a more comprehensive strategy to maintain and attract growth industries.

Notes

1. Economic base theory describes the employment multiplier, which allows economic development planners to figure the exact number of secondary jobs a particular facility in the primary or export sector will create.

2. It is interesting to note that the gap in earnings is increasingly due to the difference in paid hours worked. In 1996, manufacturing workers worked an average of 41.6 hours per week while service sector employees worked an average of 32.4 hours. Since 1988, the average weekly hours worked has grown slightly for manufacturing and fallen slightly in the service sector, adding to the discrepancy seen in wages (U.S. Department of Labor, Bureau of Labor Statistics, 2000).

3. Part of the Washington administration's industrial retention strategy was an early warning system. The Westside Jobs Network, a coalition of community and labor organizations, was funded to establish an early-warning plant-closing system. The Network roused public uproar when Hasbro-Bradley announced the closure of its Playskool facility in Chicago in September of 1984. The company had received loan funds from the City and agreed to create 400 jobs in return. In a precedent-setting move, the City sued Hasbro-Bradley for violating the agreement. A settlement was reached that required the company to retain some workers, provide financial incentives for other companies to hire laid-off workers, and to fund an emergency pool for workers. The plant eventually closed. The Playskool case is discussed in more detail in Mier and Giloth (1993).

4. These were separate departments under the Washington and Sawyer administrations. Under Mayor Richard M. Daley, the departments were folded into the Department of Planning and Development.

5. There is no written explanation for this rather unique formula. Many insiders conclude that it worked to the advantage of elected officials—residents vote, businesses don't.

6. The head tax assesses employers of over 50 employees $4.00 per employee per month. In 1996, the city raised $25.5 million from the tax, approximately 1.3% of city revenues.

7. West Town United has focused on the issue of tax reassessment with an education campaign about how it has contributed to gentrification. Further, the organization provided technical assistance to residents in filing appeals to their tax reassessments.

8. Bonds were floated for the Goose Island development (Weber, 1999).

9. Intermodal transportation centers transfer freight from one mode of transportation (e.g., rail, truck, plane, ship) to another. The most common transfer is from rail to truck. Chicago is particularly well positioned for the growth in

this industry because the railroad freight industry is growing and Chicago has six rail lines (DeVries & Lenz, 1999).

10. The Village levies a 3% tax on telecommunications services, charged on local phone bills.

5

Commercial Revitalization
in Central Cities and Older Suburbs

Local economic development planners need to be concerned about and engaged in commercial revitalization efforts because a good commercial retail base is a necessity for a strong local economy. Essentially, local economic development planners are seeking to foster what has been called the "virtuous cycle of retail" (U.S. Department of Housing and Urban Development [HUD], 1999a, p. 3). Within this cycle, businesses move into economically disadvantaged areas and, in so doing, hire local residents, leading to increases in local incomes. Higher local incomes enable residents to have higher spending and savings. Businesses, in turn, experience higher demand and resulting higher profits. They then hire more retail workers that again leads to rising local income and consumer demand.

> The cycle feeds itself. Creating sustainable economic development and allowing low-income communities to have more assets, more quality shopping, a stronger tax base, and more jobs. (HUD, 1999a)

Most retail activity is now concentrated in some form of shopping center—whether it be the large regional mall, older strip center developments, new outlet malls, or power centers. There were over 44,000 shopping centers in the United States in 1999. They contributed $47.5 billion dollars in state sales tax revenue and accounted for 8% of total U.S. nonagricultural employment (International Council of Shopping Centers

[ICSC], 2000). Thus, the importance of this sector to the local economy should not be ignored.

Several additional benefits stem from creating "virtuous cycles of retail" in central-city neighborhoods and older suburbs. First, increased proximity to necessary and desirable retail goods and services is created for residents. This promotes equity in access to good consumer choices and fair prices for goods and services. Second, there are environmental and health benefits from increasing the provision of retail goods and services within walking distance of community residents. Third, there are safety benefits to the community: The increased presence of legitimate business activities acts as a deterrent to illegal activity, vandalism, and crime.

The decline in commercial activity in central cities and older suburbs has multiple sources. The retail infrastructure of these areas is aging, or even obsolete, for today's retail requirements. This retail infrastructure is saddled with a legacy of poor design, particularly in the older suburbs. The worst of this design can be found along the retail strips found in the older suburbs as well as along key transportation corridors in the central city. The explosion of retail space in the newer suburbs is such that America has literally become "overstored." In the last decade, the lion's share of this new development has been in the form of the "Big Box." Until very recently, Big Box development's overwhelming preference has been for greenfields and land parcels that can accommodate large building footprints as well as lots of parking.

There are, however, emerging opportunities for commercial revitalization in central cities and older suburbs. The suburban market has begun to show signs of being tapped out, or overbuilt. Thus, central cities and older suburbs may be the last "frontier" for the Big Box development. Retailers are recognizing that greater population density (particularly in central cities) can compensate for lower household incomes as well as constitute sufficient market or consumer purchasing power.

The focus of this chapter on commercial revitalization for central cities and older suburbs is threefold. First, we examine the tried-and-true principles of "Main Street" revitalization for their applicability to aging suburbs and central cities alike. Next, we examine the "Big Box" retail trend and how aging suburbs and central cities can best capture these retail outlets in a manner that minimizes their negative impacts. Last, we examine the most difficult challenge of all—the need to reinvent the declining and obsolete strip mall and retail strip developments that are economic and visual blights on central cities and older suburbs.

The Main Street Approach

Aging suburbs and central cities can find valuable lessons in the economic development efforts to protect and strengthen the main streets of Small Town, USA. Shifts in retailing trends—first, the development of regional malls and, later, Big Box retail—took away business from locally owned, or older, small retail franchises typically found along these streets. Main Street revitalization efforts have focused on strengthening the local business district through careful analysis of its strengths and weaknesses and adopting a proactive approach to counter loss of business activity. At the national level, there is even a nonprofit organization specifically devoted to the preservation of main streets: the National Trust for Historic Preservation's Main Street program. This program identifies its Main Street Approach as one that

> Builds on downtown's inherent assets—rich architecture, small businesses, a connection with the past, and a sense of place—not only to develop the district as a successful market place, but also to make it the center of community identity once again. (National Trust for Historic Preservation, 2000, p. 1)

Although older suburbs, in contrast to central cities, are unlikely to have had a focal center of community identity, both types of areas should ensure that they have such a center as a key goal of their commercial revitalization strategy. Central cities can house rich architecture and historic connections.

This section interweaves a discussion of three fundamental components of a main street or commercial district revitalization strategy: market analysis and marketing for business retention and recruitment, image and design improvement, and community building. While the first two components are rather straightforward, the third component of community building requires elaboration. Community building in a successful retail revitalization strategy can require the involvement of business owners and chambers of commerce, community groups such as churches and nonprofit development organizations, as well as local government. Their efforts focus on strengthening existing businesses, identifying the needs for and bringing in new businesses, as well as making the community a more attractive and safe place for customers to come to shop.

Robert Gibbs (2000), a national retail expert, says that to be competitive, main street retailers must understand modern retail trends. For example, 65% of all sales occur at discount stores and 72% of all sales occur after 5:30 p.m., or, on Sunday (Gibbs, 2000). The average time spent in a "mall" is decreasing. Shoppers typically go to a department store at the mall, complete their business, and then go home. This, according to Gibbs (2000), is a function of people having less time to devote to shopping. Further, shoppers who used to buy in malls are now more willing to shop on "Main Street." Retailers are responding by offering more "lifestyle" retailing, and by "dechaining" (or individualizing) the chain stores. Consequently, stores are moving away from the uniform chain look, and they also want to be away from malls. For example, the home and lifestyle chain of Crate and Barrel has made the decision it will not open more stores in malls.

The stores that will attract customers, according to Gibbs (2000), have clear (not tinted) glass windows for at least 70% of their storefront. They are located close to the street—from 2 to 10 feet away from the sidewalk. They have on-street parking and operating doors every 150 feet.

Further, although retail centers should have anchors, they do not have to be the traditional anchors of a large department store or discount store. Gibbs (2000) says alternative anchors such as churches, police stations, and libraries can also work. The key is to build retail momentum around buildings that attract people who would not normally be in the area.

Strong grassroots support is emphasized by Gibbs and others as important to the retail revitalization strategy. The formation of a merchant's association can be one form of grassroots support. Loukaitou-Sideris (2000) describes the retail revitalization efforts of the Pico-Union neighborhood of Los Angeles that was a "vibrant streetcar strip" at the beginning of the last century and, subsequently, became "a poor inner-city thoroughfare with run-down establishments and boarded-up storefronts" (p. 7). Prior to the efforts of a consortium of local community institutions initiated in 1997, Loukaitou-Sideris (2000) described the rich mix of ethnic merchants as isolated from, and suspicious of, one another. The merchants did not share information with each other, collaborate, or use their collective voice to raise concerns about how issues such as crime and blight in the neighborhood were affecting them. Their isolation from each other also meant they did not recirculate local dollars by purchasing from local vendors. The efforts of the consortium brought the local business owners together to form a merchant's association. This association developed a mission statement, identified its assets and key issues, and undertook a series of community building steps to strengthen the retail sector and ties to the community. They raised money for façade improvements and holiday promotions, adopted trees and trashcans, and sponsored a local youth art program that created ceramic address markers for their storefronts.

Loukaitou-Sideris (2000) observes the following:

> Community building has been an essential part of the merchants' activities. As the merchants have become friends, advisors, and hosts to each other, they have also started to buy from each other, recycling local dollars. They have begun to understand their potential to exercise political leverage on behalf of their community. (p. 8)

Local governments have an important role to play in community building and commercial revitalization. This role can range from providing economic incentives (tax rebates, low-interest loans, and rent subsidies) to reducing visual blight (cracked sidewalks, trash and litter, design and zoning code violations) to enhancing public safety in the targeted areas. From a survey of seven southern California cities, Loukaitou-Sideris (2000) developed a series of four recommendations for commercial rehabilitation. The first calls for the stimulation of joint financing between the private and public sectors. Second, local governments should engage in outreach to let merchants know about existing commercial rehabilitation programs. The third and fourth recommendations are to encourage rehabilitation for visual leverage: that is, "in areas that draw attention to themselves because of their architecture or historic importance," and focusing on one block at a time because "having a series of rehabilitated facades provides a stronger visual impact than having them spread out along the strip" (Loukaitou-Sideris, 2000, p. 10). Loukaitou-Sideris's (2000) survey found the biggest disadvantage to their business locations was crime and the fear of crime. Local governments can help overcome this disadvantage by increasing the amount of community policing (particularly of police who walk or bike the area) to reassure customers and merchants. This can also serve as a financial benefit if it means small central-city and older-suburb small businesses do not have to pay out of pocket for private security because of a lack of public policing.

Marketing and image improvement are inextricably linked steps for the successful revitalization of a commercial district. Marketing is focused on increasing consumer awareness of, and attraction to, the commercial district and its retail outlets. To do so, the negative image that has led to decreased business and activity in the district has to be corrected.

Effective marketing must be preceded by competent marketing analysis. One focus of marketing can be to attract more businesses to the commercial district. Local economic developers and agencies can facilitate this process by determining what unmet consumer demand exists in this district. If it is determined that residents of the area regularly shop elsewhere because the desired goods and services are not available locally, then the district can target the recruitment or establishment of retail facilities providing these "outshopped" goods and services and develop a marketing strategy focusing on this goal. As an alternative, if it is determined that residents' desired goods and services are available in the retail district but retail outlets do not have enough customers, then the marketing strategy will need to focus on attracting customers back to these retail outlets. In addition, the commercial revitalization strategy may have as a goal increasing the trade area that consumers are drawn from by providing more unique destinations and retail selections. Thus, the local retail economy is strengthened by going beyond just meeting local residents' basic consumer needs to providing shopping opportunities that "import" consumers and their dollars from a larger area.

The mismatch between consumer desires and local goods and services must be analyzed to create an effective marketing strategy. The mismatch could be due to the range of goods and services exceeding or falling short of those typically consumed by the demographic and income profiles of the consumers who normally would shop in the retail district. The typical steps in a marketing analysis include the following:

- Defining the retail trade area

- Preparing a profile of the population (e.g., consumers) in the retail trade area

- Preparing a profile of the businesses in the commercial district

- Preparing a profile of existing and potential space for retail activity in the commercial district

- Preparing a SWOT (strengths, weaknesses, opportunities and threats) analysis of the commercial district

A thorough marketing analysis is created through a combination of secondary data collection and analysis and primary data collected through surveys and focus groups of the businesses in the commercial district and population in the retail trade area. Local economic developers and development organizations, depending on their resources, may elect to conduct these marketing studies themselves, or, to hire marketing consultants. Public and commercial data are available to provide the secondary data analysis needed for the marketing study. We highlight here the available public data.

The U.S. census of the population provides data needed to prepare the demographic profile of potential consumers. Included in this profile should be data on the number and kind of households in the area, age profiles, race and ethnicity, as well as income and education levels. The different lifestyles associated with different household and demographic groups translate into different retail consumption patterns. Analysis of these data enables the economic developer to determine what proportion of the area's potential consumers are, for example, young and single professionals who might be more attracted to trendy clothing stores and nighttime entertainment, retirees who might have greater needs for medicines and health supplies and more time for leisure activities, and families with children who have greater needs for toy stores and children's clothing stores. A strong ethnic presence in the population (i.e., Asian, Hispanic, Italian, Middle Eastern) can suggest stronger consumer desire and retail potential for stores and restaurants that specialize in specific ethnic cuisine and goods. The consumer demand for these can be provided in specialized stores, or, in expanding the selection of goods in traditional stores.

Data on how the population spends its income can be obtained from the Survey of Consumer Expenditures, conducted by the Bureau of Labor Statistics within the U.S. Department of Labor. Combining these data with the population data described previously, the local economic developer can make an estimate of how much local consumers spend on different retail categories: clothing, food in the home, food eaten away from the home, books, entertainment, auto services, entertainment, furniture and electronics, and so forth.

These data can be compared with estimates of the sales volume different local businesses have in the area: grocery stores, gasoline stations, restaurants and bars, apparel stores, bookstores, movie theaters, and so on. Data on business sales can be estimated from the Survey of Retail Expenditures conducted by the U.S. Department of Commerce.

By comparing the estimates of consumer purchases in the retail trade area with those of business sales in the commercial district, the local economic developer can develop an estimate of the unmet retail potential of the commercial district. This analysis can then be used to target specific questions for customer surveys or focus groups to help the local economic developer gain a more detailed and accurate understanding of the area's redevelopment potential on which to base the goals and objectives of the revitalization plan. For example, participants in the surveys and focus groups can be asked what goods and services they would like in the area, why they shop outside the area, and what their image is of the area.

The market analysis allows the local economic development planner and community to develop a realistic strategy for commercial revitalization. A critical component of implementing the revitalization strategy is marketing to potential new businesses and to new and existing customers. The marketing plan must address the particular image or vision the retail area wants to project for itself. An important outcome of the efforts of this plan should be correcting the negative image associated with the declining commercial district. As one marketing handbook states, "It is important to understand that the power of image is working all the time" influencing people's actions and reactions. When those responsible for the commercial district or area do not pay attention to its image, they give up their role in guiding the public's response and leave the public with a continued negative image ("Mainstreet," 1999).

A number of aspects of image must be addressed for commercial revitalization. First, is the visual image of the district itself. Are the buildings run down? Are the streets and sidewalks in disrepair? Are graffiti and trash a problem? Are businesses storing abandoned equipment on-site? A commercial district with good design standards and code enforcements for its buildings and infrastructure will generate a far more positive image than one without them. A component of the revitalization strategy will need to address these issues and develop strategies to improve the physical character of the area. Examples of efforts to do so are highlighted in this chapter's section on "Reinventing the Strip."

Big Box Coping Strategies

> Much as regional shopping malls displaced traditional downtowns during the sixties and seventies as the retail centers of cities, discount superstores today challenge the older malls as well as the downtowns struggling to recover from the effects of malls built a generation ago. (Beaumont, 1997, p. 3)

The "Big Box retailer" or "category killer" store that has now become the dominant form of all new retail construction can range in size from 90,000 to 250,000 square feet. It is typically accompanied by a vast parking lot that can be "as big as ten or more football fields" (Beaumont, 1997, p. 3). The prototypical Big Box store is just that: a Big Box. It is a plain, windowless one-story building surrounded by asphalt.

Beaumont (1997) aptly describes this retail form:

> The biggest superstores include such discount department stores as Wal-Mart, Target, Kmart, Shopko, and Meijer's. These stores sell a wide variety of products ranging from clothing to cameras to garden supplies. Examples of warehouse clubs include Price-Costco and Sam's Club. These stores offer a more limited menu of goods at wholesale prices. Stores like Home Depot, Toys R Us, and Staples are known as "category killers." They specialize in selling certain types of products, such as home improvement, toys, or office supplies. Retail analysts say these stores often kill off the competition in their specialty; hence the name "category killers." "Power centers" are shopping areas in which superstores are clustered. (p. 4)

The Big Box has become the dominant form of retail because it is preferred by consumers and financers alike. Institutional investors and other lenders favor Big Box developments over conventional retail tenants because the Big Box firms typically have good corporate credit ratings (New Jersey Office of State Planning, 1995). Consumers are preferencing Big Box retail because of the greater merchandise selection they offer, the greater convenience they offer that is prized particularly by two-income households, and because they are perceived to offer lower prices and greater value (New Jersey Office of State Planning, 1995).

The Big Boxes were originally introduced into new suburbs and, in rural areas, typically between small towns (the Wal-Mart location strategy). However, as the retail markets of these areas became tapped out, Big Boxes began to turn their sights on central cities and inner-ring suburbs. There has been genuine and legitimate concern raised about the displacement effects and negative impacts such Big Boxes have had on the main street businesses of small towns, and for their contribution to the sprawl pattern characterizing all of the nation's metro areas. Yet the redirection of Big Box attention to central cities and inner-ring suburbs *can* be a positive economic development trend. This is because despite the fact that the nation as a whole is overretailed, inner cities and inner-ring suburbs are often underretailed. As a result, the residents of these areas do not have equal access to the range or quality of competitively priced goods that their suburban and small-town counterparts have. Moreover, this means retail employment, an important sector for lower-skilled workers, is less available to central-city and older-suburb residents. We focus in this section on the lessons that economic development planners for central-city and older suburbs can learn from the experiences and responses small towns and new suburbs had as the original recipients of Big Box development.

First, however, we review recent statistics revealing "the untapped potential" of urban markets that has begun to draw the attention of Big Box and other retailers to inner cities and inner-ring suburbs. The U.S. Department of Housing and Urban Development ([HUD]; 1999a) has identified large retail gaps existing in the nation's cities whereby cities do not capture adequate proportions of their own residents' buying power or that of visitors to them. This gap has developed because of the decades-long commercial and retail investment in suburban greenfields. As a consequence, inner cities and inner-ring suburbs are "understored."

This recent HUD (1999a) study focusing on central-city neighborhoods estimated their retail purchasing power at $331 billion in 1998. However, because these areas were understored, they experienced significant retail gaps. For large cities such as New York and Los Angeles, these gaps were estimated to be as large as $119 billion and $58 billion, respectively. Medium-sized and smaller cities also had significant gaps: Schenectady, New York's was $701 million, while Charleston, South Carolina's was $1 billion (HUD, 1999a).

Beyond the current untapped demand, HUD notes that the density of demand found in urban areas offers an additional benefit to retailers. That is, the higher population density of urban areas can more than compensate for the higher average household incomes found in low-density suburban areas. For example, HUD (1999a) found the following:

> Retail demand per square mile in inner-city Boston is six times as great as in the Boston metro area as a whole. Not surprisingly, the Super Stop and Shop store in Boston's inner city is the highest grossing of that company's 186 supermarkets. (p. vi)

Oakland, California, has a Kmart SuperK store with sales that are 50% greater than those of the chain's comparable stores (HUD, 1999a).

In addition, the City of Pasadena, California, waged a successful campaign to convince the Big Box retailer, Target, to locate in its downtown in 1994. Target national sales averaged $230 per square foot in 1995. However, the new Pasadena store's sales averaged $280 to $325 per square foot, or 30% to 35% higher than anticipated, for the same year (Beaumont, 1997).

When there is inadequate retail or understoring in inner-ring suburbs and low-income inner-city neighborhoods, residents "outshop" for their needed goods and services. For the inner-city neighborhoods alone, HUD (1999a) estimates this outshopping represents billions of dollars annually. Economic development planners need to help these areas capture more of their residents' retail purchasing, convincing businesses there is more than adequate consumer demand for these stores as well as pools of untapped labor to staff them.

In doing so, however, economic development planners must meet an additional challenge of ensuring that these Big Box retail outlets are a good fit with the scale and design of the existing neighborhoods in which they locate. Fortunately, recent experiences across the nation suggest *determined* communities can successfully negotiate with Big Boxes to obtain stores that not only are a better fit than the typical one-size- (and style) fits-all approach but have a greater chance of successful reuse if (but more accurately when) the Big Box retailer moves on. In fact, it has been estimated the average life of a Big Box retail outlet is only 15 years. Wal-Mart, the retailer that started the Big Box trend, had 333 empty buildings distributed across 31 states, comprising 10.5% of the stores it owns or leases (Norman, 1999). Furthermore, 15 of these states had 10 or more empty Wal-Mart stores.

While inner cities are understored, America as a whole (but read suburban America) is overstored. This overstoring has resulted in stagnant sales per retail outlet. However, the industry response has been to open more new stores, in particular, superstore versions, to compensate for the inadequate sales performance in existing outlets (Rossi, 1998). This is occurring across all categories of retail: groceries, drugstores, hardware and home improvement, toys, sports, clothing, books, and so on. In fact, we now have 20 square feet of retail space per person in the United States, up from 15 square feet in 1986 (Norman, 1999).

In the overstored American retail landscape, retailers have begun to focus on what might be called "the last frontier," that is, understored inner cities and older suburbs (Bressi, 1996). This can be a positive movement for these areas because, as we previously noted, they lack adequate retail opportunities in their own areas—opportunities to buy and to work. However, the greenfield suburban experiences with Big Box retail over the last two decades have generated community backlash, acting as a catalyst for efforts to control sprawl development.

Beaumont (1997) groups the problems associated with "sprawl-type superstores" into four categories: design, environmental, auto-dominance, and fiscal-economic. From a design perspective, these superstores are undesirable because they are built as one-story, inward-looking or "fortresslike" buildings set at the back of sites with acres of parking in front. Their sites lack trees, landscaping, and amenities such as bus shelters and benches. They have very large signs, typically mounted on tall poles for self-advertisement. Their design does not reference any relationship to the community nor does it seek to preserve and reinforce local character (Beaumont, 1997).

From an environmental perspective, sprawl-type Big Boxes lead to the unnecessary destruction of greenfields and trees, contribute to increased air pollution, nonpoint source water pollution, and storm-water runoff problems. The unlandscaped parking lots become heat islands (Beaumont, 1997).

Beaumont (1997) also criticizes these stores for their contribution to excessive auto-dominance. They are pedestrian unfriendly, lacking sidewalks, crosswalks, or traffic signals to facilitate safe street crossing. They do not provide access to users of public transit. Consequently, Big Boxes are inaccessible for those who either cannot afford or are unable to drive (elderly, low-income, young), and they can generate up to 10,000 car trips a day, overwhelming the surrounding community.

Finally, from a fiscal and economic standpoint, sprawl-type superstores require taxpayer subsidization for more and bigger roads to accommodate large traffic volumes, and for water and sewer line extensions to previously undeveloped lands. Further, these stores can create a retail glut if the local economy is not growing sufficiently, thereby causing displacement of existing businesses and disinvestment of existing commercial buildings (Beaumont, 1997).

Although, as Beaumont (1997) observes, "Many public officials actually think they have no option other than to accept 'Big Box' sprawl or lose the hoped-for tax revenues and jobs to a neighboring community" (p. 35), there are increasing numbers of communities proving this can be otherwise. They are employing planning tools to achieve Big Box developments that fit better with their community, minimize negative environmental and transportation impacts, and even reuse existing properties. Particularly relevant tools employed for central cities and inner-ring suburbs include retail square-footage or size caps, and design standards.

The design standards established by Ft. Collins, Colorado, have been widely cited as a model for ensuring that community identity does not yield to corporate identity (see, e.g., Jossi, 1998). Beaumont (1997) summarizes these guidelines, which apply to retail stores in excess of 25,000 square feet:

- Long blank walls that discourage pedestrian activity are prohibited.

- Building facades must be broken up with recesses.

- Ground-floor façades must have arcades, display windows, awnings, or some other feature to add visual interest to the structure.

- Stores must be accessible to pedestrians and bicyclists. They must have several entrances to reduce walking distances from cars where stores border two or more public streets.

- Stores must provide amenities, such as patio seating areas, kiosks, or fountains.

- No more than half of the store's parking may be located between the store's front façade and the abutting street.

- Sidewalks linking stores to transit stops, street crossings, and building entrances must be provided. Sidewalks must be landscaped.

- Stores must have clearly defined entrances, with canopies, porticos, or arches. (p. 41)

The wave of Big Box retailers began with the large general discount retailers (Wal-Mart, Kmart, Target) and home building suppliers (Home Depot and Lowe's). Most recently, chain drugstores have begun to turn their attention to the last frontier of central cities and older suburbs. As Stillman (1999) notes, the proliferation of chain drugstores, particularly, Walgreen, Rite Aid, CVS, and Eckerd, has displaced the independently owned stores and tended to impose "large, single-story, characterless architecture and oversized parking lots" (p. 3). Stillman (1999) states that "Communities need to be prepared to negotiate with the developers and chains and, if possible, arm themselves with regulations and ordinances that will protect their character and identity" (p. 5).

When communities have been prepared, we can find successful examples of chain stores responding with more sensitive designs and even placing their stores in historic buildings. For a historic building to be feasible, however, if must have a ground floor with the 10,000 to 14,000 square feet normally associated with standardized chain drugstore layouts. Stillman (1999) suggests the reuse of historical buildings happens more often in urban areas than in small towns where higher volumes of foot traffic remove the need for parking lots and drive-through windows.

Since the introduction of the Big Box, retailers have increased the size of their boxes. For example, Wal-Mart has moved to a supercenter concept. In essence, the first-generation Big Boxes became functionally obsolete before they become physically obsolete (New Jersey Office of State Planning, 1995). The outmoded store units have to be disposed of or sublet for different uses. Since the Big Box retail outlet inevitably becomes obsolete and abandoned by the retail chain, it is important for planners to consider future retrofitting options when evaluating the proposed site layout of the Big Box. The New Jersey Office of State Planning (1995) suggests the following: "Parking lots may become public plazas or open spaces; retail buildings may become housing, offices or civic uses. And the circulation system should be planned to facilitate future connections—including pedestrian and bicycle links—to surrounding uses" (p. 7).

There are real estate specialists for disposing of Big Box retailing, although Wal-Mart itself handles disposition in-house. One of Wal-Mart's older outlets was converted to a Mercedes dealership in Pompano Beach, California—a rather ironic move from the extreme discount end of the retail spectrum to the most luxurious end. Real estate brokers have found that the large parking lots of obsolete Big Box stores are attractive not only to car dealerships but to new users as disparate as post offices and churches. Industrial and offices users have also taken over the early obsolete Big Boxes. It is often possible to subdivide the Big Boxes into smaller retail outlets (Hazel, 2000b).

The approach to building a Wal-Mart in Lawrence, Kansas, in the early 1990s offers local economic developers an important example of planning for the reuse of such facilities. The store was designed to be converted to affordable housing when Wal-Mart no longer needs the operation. This was done by increasing the standard height of the Wal-Mart from the usual 14 feet to almost 20 feet to allow for the addition of a second floor and conversion of space into apartments. In addition, the wall coursing was detailed in a manner that allows for the removal of concrete blocks and replacement by windows in the apartment conversion process (Wal-Marts are typically designed with no windows [Greening Wal-Mart, 1993]). While the 120,162-square-foot building still looks like a typical Wal-Mart, it also contains a number of building and design innovations to minimize its impact on the environment. Among these were creating the roof and ceiling out of an engineered wood beam system instead of the usual steel; using a skylight system that allows 40% more daylight into the building and thus reduces energy use; and landscaping irrigation partially provided by recovered water from rest room sinks and drinking fountains. Moreover, the parking lot was made from recycled asphalt, and parking lot signs and bumpers were made from recycled plastic (Wilson, 1993).

In the latest round of evolution, the Big Boxes are discovering smaller stores are not really so undesirable. Best Buy, Circuit City, and Home Depot are some of the Big Box retailers that have been introducing small-format stores aimed at smaller markets such as towns with populations of 100,000 to 200,000 or even neighborhoods. Jim McCartney (2000) writes the following:

> Ultimately, the loser in this trend will be the small-town merchant or the neighborhood store...The Home Depots and Wal-Marts can negotiate lower prices on the products they purchase, and they don't have to pay a middle man—such as a wholesaler. Their massive, yet efficient, distribution system allows them to keep their stores full and their inventory and shipping costs low. (para. 8)

To help the small-town and neighborhood merchants respond, local economic development planners will need to engage in market analyses and niche identification for these retailers. They will be able to compete best with unique product offerings and superior customer service.[1]

Communities have been confronted with many challenges as the face of retailing changed and Big Boxes inserted themselves into their commercial landscape. This section outlined a number of approaches to coping with Big Boxes that allow communities to mitigate the negative impacts and increase the benefits of becoming home to such retail forms. A final "hardball" approach to coping with the mobility of Big Boxes is for the community to require the Big Box retailer to put up a performance bond (Peirce, 1996). The bond becomes payable if the Big Box does not stay at the location for a long enough period of time to pay off the public improvement costs and does not find a viable user for the facility when it leaves. The bond can be used to pay the costs of converting the facility for another use, or, for disposing of it.

Reinventing the Strip and Its Retail Centers

> Changing consumer shopping patterns do not bode well for smaller strip centers. That's because retailers and grocery store operators are offering "one-stop shopping" by sandwiching everything from a bank office to a fast-food restaurant inside their stores. (Johnson, 1996, p. 1)

It is difficult to estimate how many obsolete or subperforming strip shopping centers there are, but few cities or towns in the United States are not plagued by these old, unattractive, and dysfunctional retail centers.

Strip shopping centers are distinguished from shopping malls in that malls are enclosed structures with parking on the outside, whereas strip centers have stores that open onto a parking lot (Colman, 1992). They are typically sited on strip thoroughfares where there is a hodgepodge of commercial signs, lots of vacant stores, ill-maintained parking lots, and frequent curb cuts into traffic that contribute to its stop-and-go nature. These strip thoroughfares are pedestrian unfriendly, lacking sidewalks and crosswalks at intersections. Moreover, they lack any design theme, streetscaping, landscaping, or all three. Based on the retail center types in Box 5.1, if we define strip centers as neighborhood centers or community centers, then there are over 40,000 potential strip centers in the United States.[2]

The preponderance of obsolete strip centers are found in the central city and in older suburbs along major corridors. They contribute to economic and visual blight, and represent foregone opportunities for business and job creation, as well as income and tax revenue generation. Strip shopping centers arguably pose the greatest challenges to communities seeking to redevelop and maintain a strong retail sector.

Box 5.1
Types of Retail Centers

- *Neighborhood:* Convenience goods and personal services for daily life, usually anchored by supermarket; 30,000 to 100,000 square feet

- *Community:* Provide a wider range of stores than neighborhood center; usually anchored by small department store, variety or discount department store plus supermarket; 100,000 to 450,000 square feet

- *Power:* Anchored by a market dominator such as Toys R Us and discount department store or warehouse club; at least 350,000 square feet

- *Specialty:* Has a dominant marketing theme such as home improvement, entertainment, factory outlet, etc.

SOURCE: Johnson (1996).

The final section of this chapter focuses on efforts to revitalize "the strip" and its retail centers. It focuses, in particular, on the efforts of a community development organization in an older, unincorporated suburb of the Atlanta metropolitan area.

The Urban Land Institute (ULI), based on an analysis of ailing commercial strips in the Washington, D.C., area, has developed a set of basic principles for reinventing suburban strip centers (Gentry, 2000). Those who are setting out to redevelop or reinvent a strip center should begin with the development of a marketing plan that fosters communication between the public and private sectors and includes area residents. This plan should reflect expected changes in demographics and retail demand so the short- and long-term needs of the area can be addressed.

Strip reinventors should determine whether there is a need to prune the retail-zoned land, as a surplus of such land can facilitate the abandonment of existing retail buildings and centers and encourage construction of new retail space. This raises vacancy rates in the area at the same time it contributes to further sprawl. ULI also recommends the redevelopment plan create specific focal points of intense development that are connected by less developed areas, open space along the strip, or both, a concept they refer to as "pulsing" development. Further, traffic needs to be tamed or calmed. This means roads should not serve as edges dividing development (which encourages faster driving) but rather as seams to connect different portions of the development. Pedestrian routes should be planned in a manner that minimizes conflict with vehicular movement. Strip reinventors can ameliorate the ugliness of auto-dependent landscapes by designing wide median strips with mature trees and good lighting, as well as well-landscaped parking lots and areas. They should strive to create a sense of community that is safe, comfortable, and attractive. By diversifying the development so there are a variety of uses (not just retail), they can foster greater use of the area. Lastly, ULI recommends public officials lend their support to funding and zoning decisions that complement, not inhibit, strip redevelopments. In conjunction, they should consider placing public investments such as government agencies or post offices into redeveloping strip centers to promote mixed-use development.

Ferguson, Miller, and Liston (1996) recommend establishing a vacant lot and storefront program that surveys owners of such properties and surveys residential and business neighbors to determine what

Figure 5.1.   Typical Block of Strip Development Along Roswell Road
SOURCE: Reprinted with permission from Sandy Springs Revitalization, Inc. These
          projects were made possible in part by funding from Fulton County, Georgia.

they would like to see in such properties. The program cleans up vacant
lot and storefronts and seeks to get abandoned property resold and put
into viable use. It also explores nonretail businesses when retail is no
longer feasible.

Marketing of strip centers has always been challenging. These centers typically do not have a common area to stage public-drawing events
nor can they afford to hire a marketing or even a general manager who

has a discretionary fund to advertise the centers. Hazel (2000a), in *Shopping Centers Today*, writes the following:

> Getting strip center tenants to fund any kind of program is increasingly difficult: The supermarkets and discount store that typically anchor small centers do their own advertising and won't contribute, while the mom-and-pops rarely wish to spend the money. (para. 11)

Tenants in a strip center have to join together to develop a marketing plan. Increasingly, those strip centers that engage in marketing are using direct mail or the Internet. Traditional marketing in the form of advertising in local newspapers and on local radio stations has been found to be ineffective because their reach is too broad. Direct-mail coupons and gifts with purchase are proving more effective in getting local residents to come to the strip center. The Internet is now being used by Kimco Realty Trust, the nation's largest owner and manager of neighborhood centers. Kimco provides a center Web site with a page for each merchant that describes store locations and hours (Hazel, 2000a).

Reinventing Sandy Springs Strip[3]

Sandy Springs, an older-suburban community in metropolitan Atlanta, is one of the nation's largest unincorporated areas. It is home to over 80,000 people and encompasses 35 square miles. The major north-south route, and the community focus of Sandy Springs, is Roswell Road. This road is a particularly undesirable form of the older, urban strip development found throughout the country. In its own revitalization plan, Roswell Road is described as a "chaotic commercial strip. . .[with] different existing urban design and visual characteristics from one end to the other" (HOH Associates, 1993, p. 1.11).

At its southern end, the land uses are primarily office and residential and there is heavy vegetation. The northern one half of the Roswell strip is characterized by a high degree of competing signage and aboveground transmission and utility lines. The building densities and setbacks vary widely as one moves along the road. Traffic congestion during the commute hours and the lunch hour is extreme. Approximately 40,000 vehicles travel the road each day. A major hindrance to the traffic flow stems from the fact that nearly every single tenant store and shopping center along the road has one or more curb cuts that allow for turning in different directions.

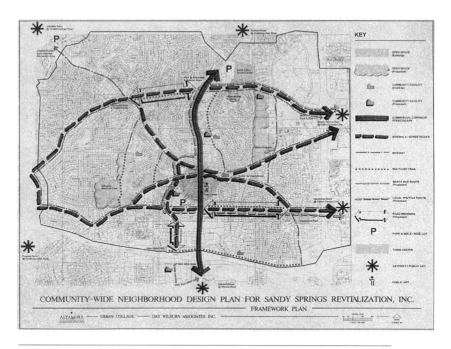

COMMUNITY-WIDE NEIGHBORHOOD DESIGN PLAN FOR SANDY SPRINGS REVITALIZATION, INC.
FRAMEWORK PLAN

Figure 5.2.   Community-Wide Neighborhood Design Plan for Sandy
Springs Revitalization, Inc.

SOURCE: Reprinted with permission from Sandy Springs Revitalization, Inc. These
projects were made possible in part by funding from Fulton County, Georgia.

In 1993, the Sandy Springs community, assisted by a consulting firm, created an urban design concept for Roswell Road that seeks to upgrade its image, function, and character. The road is to be divided into three types of zones with design treatments created that reflect the character, density, and scale of each. The zones are (1) northern and southern commercial strip, (2) suburban corridor, and (3) main street redevelopment zone (see Figure 5.2.).

The Sandy Springs Revitalization Plan led to the chartering in 1994 of a private, nonprofit community-based planning and development organization, Sandy Springs Revitalization, Inc. (SSRI). SSRI is funded through service contracts with the county, membership fees, contributions, and grants. A key goal of SSRI is the creation of a community

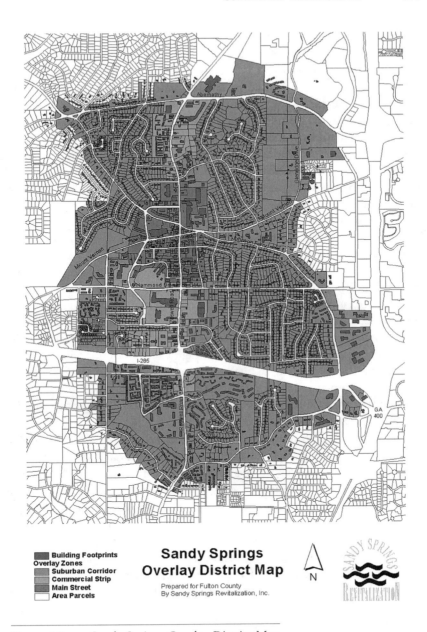

Building Footprints
Overlay Zones
Suburban Corridor
Commercial Strip
Main Street
Area Parcels

Sandy Springs
Overlay District Map

Prepared for Fulton County
By Sandy Springs Revitalization, Inc.

N

Figure 5.3.   Sandy Springs Overlay District Map

SOURCE: Reprinted with permission from Sandy Springs Revitalization, Inc. These
         projects were made possible in part by funding from Fulton County, Georgia.

focus and physical identity for the community's commercial districts, especially along Roswell Road.

The plans for Roswell Road were further solidified by SSRI in 1997 in a structured public participation process that focused on the relationship of the Roswell Road commercial corridor and its adjacent residential neighborhoods (Sandy Springs Revitalization, Inc. [SSRI], 1997). The process yielded a plan structured around three concepts: destinations, circulation systems, and urban design. Destinations are community facilities and open spaces (public and private) around which the plan will create linkages between the residential neighborhoods and the commercial corridor. The circulation systems detailed in the plan will increase accessibility for pedestrians and cyclists. To accomplish this, sidewalks, bikeways, and multiuse trails on Sandy Springs major arterials are planned. New parking facilities are called for at strategic locations to promote pedestrian-friendly retail development along the commercial corridor. Finally, a local shuttle system is planned to serve the Sandy Springs commercial district. The urban design elements of the plan seek to improve visual character and facilitate movement between neighborhoods and the Roswell commercial strip. The plan proposes design guidelines for commercial signage, traffic-calming measures, as well as enhancements to signal neighborhood entranceways. To strengthen the identity of the community, gateway treatments and public art installations are called for at community boundaries.

The primary vehicle for pursuing the goals of the revitalization plan is the overlay zoning district and development standards Sandy Springs proposed and received approval from the county to implement. The overlay district is a planning tool used when local government's general zoning and development standards do not address a subarea's unique issues and conditions. In the case of Sandy Springs, the county's commercial zoning categories were designed for lower-density suburban areas than Sandy Springs had become.

The overlay area divides Sandy Springs into three districts that have slightly different changes to the general zoning guidelines. The overlay plan seeks to foster or create a Main Street District out of a portion of the current Roswell Road urban strip corridor. It requires regularly spaced street trees on both sides of the road, decreases the requirements for building setbacks, and increases the height limits for buildings. Sidewalk widths are increased and freestanding signs are eliminated on properties that use the reduced setback. The changes in standards seek to create the greater density and pedestrian-friendliness of the traditional downtown.

The overlay district took SSRI 3 years of consensus building to be accepted by the community and, thus, approved by the county commission (J. Cheek, personal interview, November 3, 2000). The overlay standards apply to property owners and existing businesses only when they seek a sign permit, building permit, or land disturbance permit. The standards apply for all new construction and for site re-development when the declared value is equal to, or greater than, 60% of its appraised value. The overlay district removes the county require-ment of acceleration-deceleration lanes for individual businesses along Roswell Road that ran counter to the goals of creating a pedestrian-friendly environment.

SSRI, in an innovative approach, has served as the design-build contractor for Fulton County on certain commercial streetscape and neighborhood design projects for the Sandy Springs revitalization plan. It began with one block of commercial streetscape on the west side of Roswell Road to demonstrate what the Main Street District streetscape would look like when the *Sandy Springs Revitalization Plan* was completed.

This one-block segment features distinctive green street lamps, 9-foot sidewalks with a 2-foot brick separator strip, and a canopy of street trees and landscaping with row hedges to screen parking that fronts the street. The pedestrian amenities it adds to Roswell Road are benches, trash receptacles, and planters. A new Kinko Copy Center building was built closer to the street, with parking in the rear. This "demonstration" project facilitates the implementation of the revital-ization plan by creating great public interest and serving as a model for subsequent private projects. (See Figure 5.7.)

A key component in the successful implementation of the overlay district is the Sandy Springs Design Review Board. This board is made up of Sandy Springs residents, property owners, and business owners. It makes nonbinding recommendations to the Fulton County Board of Zoning Appeals for projects that require certain kinds of permits. The Board of Zoning Appeals has yet to go against a recommendation made by the Design Review Board. If the Design Review Board did not exist, SSRI would have to rely entirely on the County to implement its overlay standards. This would greatly increase the difficulty of implementing the

Figure 5.4.   Goodyear Tire Retail Outlet Before Renovations
SOURCE: Reprinted with permission from Sandy Springs Revitalization, Inc. These
        projects were made possible in part by funding from Fulton County, Georgia.

overlay district because its standards and regulations are so different
from the County's established procedures. An additional benefit of
involving the Sandy Springs Design Review Board in the implementation
of the overlay district is it performs the role of enforcer of the ordinance
allowing SSRI to be seen in the more positive, less controversial role of
economic developer and project implementer (A. Steinbeck, personal
correspondence, April 4, 2001).

Since it was adopted, SSRI has found the need to "tweak" some of
the requirements of the overlay district to make it more effective. For
example, as of this writing, it was seeking to change the standard parking requirement from a minimum to a maximum to decrease the amount
of available land for parking and increase the amount for leasable building space. The standard parking requirement is five parking spaces for
1,000 square feet of retail space, and three spaces for 1,000 square feet
of office space. Further, it seeks to change the requirement that overlay
district standards apply to redevelopment of a site from when the proposed redevelopment will equal 60% to 40% of the declared value.
Hence, more properties will conform to the standards sooner. It also

Figure 5.5. Goodyear Tire Outlet After Renovations Conforming to the Design Standards of the Overlay Planning District

SOURCE: Reprinted with permission from Sandy Springs Revitalization, Inc. These projects were made possible in part by funding from Fulton County, Georgia.

seeks to create a front-door orientation to the street for newly constructed buildings in the Main Street Zone. Thus, the principal business entrance will be required to face the street, and there will be pedestrian access directly from the sidewalk to this entrance. The overlay district is also being expanded to include commercial areas farther up and down Roswell Road along with several neighborhoods.

Because creating the appropriate physical environment is only one component of commercial revitalization, the SSRI founded the Sandy Springs Business Association (SSBA) in 1998 to promote the area's economy. Within 2 years of formation, SSBA had close to 250 members. It initiated a business district marketing study. SSBA compiled and printed, for broad distribution, a membership directory that features categorical listings and display ads of the area's businesses. It formed a Retail Roundtable that holds monthly brainstorming sessions of retailers and restaurateurs for networking and developing ideas to improve the retail climate. SSBA also leverages volunteer time and money to promote the

Figure 5.6. Main Street Transformation of One Block of the Roswell
Road Strip

SOURCE: Reprinted with permission from Sandy Springs Revitalization, Inc. These
projects were made possible in part by funding from Fulton County, Georgia.

Figure 5.7.   The New Kinko Copy Center: SSRI's Model for Future
                Development
SOURCE: Reprinted with permission from Sandy Springs Revitalization, Inc. These
        projects were made possible in part by funding from Fulton County, Georgia.

Revitalization Plan. Creating a Business District Plan is the next step for Sandy Springs' revitalization efforts. This plan will seek to address the high vacancy rate of the properties within the business district. There are 30 properties overall, most are small shopping centers. Twelve of these have vacancy rates of more than 20%, and several of these have as high as 50% vacancy rates. SSRI hopes to have the business district designated a tax allocation district (the Georgia version of a tax incremental finance district) to provide funds for redevelopment.

There are significant similarities between the commercial revitalization plan Sandy Springs has developed for its strip corridor and the ULI recommendations profiled earlier for reinventing the strip. SSRI's plan creates focal points of development, tames traffic, enhances pedestrian access, and incorporates attractive lighting and treescaping. Sandy Springs began its reinvention efforts many years before ULI published its recommendations. What the ULI recommendations do not indicate, but this examination of Sandy Springs efforts suggests, is that strip reinvention is by no means a fast or simple process. It is, instead, a Herculean undertaking that requires great persistence, the formation of the appropriate redevelopment vehicles, successful forging of public and private partnerships, strong community engagement in the visioning and planning process, and organization and capacity building of the existing merchant base.

Lessons for Economic Development Practice

A good commercial retail base is a necessity for a strong local economy. Local economic development planners, who successfully create "virtuous cycles of retail" in central-city neighborhoods and older suburbs, increase their population's access to necessary and desired goods and services at fair prices. They also provide increased employment opportunities for local residents.

Commercial revitalization efforts must begin with market analysis that assesses what retail goods and services the community needs and which businesses can operate successfully in the community. Grassroots support and community building are essential to the successful implementation of community revitalization efforts. They must occur among the existing merchants of the area as well as the rest of the community (local government, community-based nonprofits, churches, etc.) that have a vested interest in the area. Marketing and image improvement are important components of commercial revitalization. These components are shaped by an understanding of existing consumer and business attitudes. At the same time, successful implementation of these components creates greater interest and positive perceptions of the district among consumers and businesses.

Developing a good commercial retail base may require retrofitting central-city neighborhoods and older suburbs with Main Streets and commercial business districts. This, in turn, provides the opportunity to create a focal center of identity, a feature especially lacking in older suburbs.

It may not be possible to immediately attract a traditional retail anchor such as a large department or discount store to the commercial district, but local economic development planners can seek out alternative anchors such as government service offices, community cultural-recreation centers, or even churches to bring activity back to the area and to support smaller businesses.

On the other hand, local economic development planners may find that there is "untapped retail potential" in their community that has caught the attention of Big Box retailers. If so, the economic development planner should help the community consider carefully the scale and design of the Big Boxes. Determined communities can be successful in negotiating designs more appropriately scaled to the community and fitting better with existing design themes. They may be able to get the Big Box to reuse an existing structure or to build a new structure that is more environmentally and pedestrian friendly. The new structure can even be built in a manner that anticipates its obsolescence as a retail outlet and accommodates alternative uses such as multifamily housing.

It is possible to reinvent retail strip development so that it takes on Main Street characteristics and serves as a community destination rather than a drive-through zone. To do so, the local economic development planner may need to go beyond traditional zoning and land use tools to the concept of the planned overlay district. This district requires new and renovating businesses to follow design standards that increase the attractiveness of the area and orients retail activity to the street and the pedestrian rather than the automobile. A strong vision and dedicated community builders are needed to carry out this particularly challenging form of commercial (and community) revitalization.

Notes

1. See *Up Against the Wal-Marts: How Your Business Can Prosper in the Shadow of the Retail Giants (1996)*, by Don Taylor and Jeanne Smalling Archer, for detailed strategies to help the small retailer survive.

2. This estimate was made based on adding the 1999 figures for shopping centers by three size categories (less than 100,001 sq. ft., 100,001 to 200,000 sq. ft., and 200,001 to 400,000 sq. ft.) published in ICSC (2000), International Council of Shopping Centers, *Scope 2000*.

3. The author thanks Sandy Springs Revitalization Director John Cheek, and Planner, Allen Steinbeck, along with Sandy Springs Business Association Director, Donna Gathers, for their contributions to this section.

6

The Reuse of Office and
Industrial Property in City and Suburb

There has been little attention given to economic and metropolitan restructuring's impact on changes in land use demands and the context for the reuse of office and industrial properties. Because of recent patterns of economic restructuring, there has been an increased emphasis on creating a more flexible organization of manufacturing production, and greater flexibility in the delivery of advanced services activity such as financial and legal services. In combination with long-term trends in the suburbanization of population and business activity, as well as the more recent changes in technology requirements, which preference new office and industrial facilities over old, demand has declined for both the large, old industrial plants and warehouses and for the older office buildings and corporate headquarters of the inner city and older suburbs. The resulting secular trends toward high vacancy rates in older industrial and office properties are of serious concern, not only because they imply underused land, buildings, and infrastructure, along with lost business and employment, but because they are also precursors to abandonment. Abandoned properties mean lost tax revenue, increased vandalism, and illicit activity. This, in turn, fuels further decline in the central cities and older suburbs where the properties are located, making the work of local economic development all the more difficult.

Local economic development planners and practitioners have a strong stake in facilitating the reuse of their communities' office and industrial properties, and they have motivated partners in the real estate brokers, financial lenders, and owners of these properties. Essential for a proactive plan to ensure the properties' reuse is an understanding of the larger economic forces that are changing the demand for them, renovation possibilities and limitations of the properties, and potential need for zoning and land use regulations changes and infrastructure modifications. We begin the next section with a discussion of office markets and the reuse of office properties. The second half of this chapter focuses on the industrial property market and the reuse of industrial and warehouse facilities.

The Reuse of Office Properties

Office sector land use is an essential component of a healthy urban economy. It provides high employment densities that in turn facilitate the efficient provision of urban services such as mass transit. Increasingly, these services are just as needed in suburbs as well as in central cities. Office sector employment typically encompasses a range or hierarchy of occupational levels from low-level clerical to upper-level managers and executives. Thus, it helps to meet the needs of a cross section of the labor force. In particular, it is an important source of employment for inner-city and older-suburb residents who have lower education levels, income, and mobility. Beyond this "captive" labor market, for cities and older suburbs that seek to promote a stronger middle class, professional workers occupying office buildings are an important group to "capture" for middle-income housing development. Last, but not least, office sector land use creates demand for ancillary business support services as well as retail services that, in turn, generates additional employment and revenue for the urban economy.

To understand trends in the office market, it is necessary to recognize its heterogeneous nature: distinctions in age, building services, building materials, and location are used to assign buildings to Class A, B, or C status. Vacancy rates among the different classes of office buildings typically vary but are linked. For example, overbuilding of the Class A office market in major cities during the speculative frenzy of the 1980s resulted in a general decline in building occupancy rates. There was a trend toward vacating older Class B and C buildings for new Class A structures, either in the central business district or in suburban office parks on the metropolitan fringe.

Defining whether a building is Class A or Class B is not an exact science. While Class A buildings are more likely to be newer construction, a new building can be built to Class B standards, and this appears to occur more typically in the suburbs. Typically, a Class A building is characterized by a prime location, high-quality finish, and high level of maintenance. It commands a high rental rate relative to its market area. Class B buildings are characterized by quality finish and maintenance, lower levels of amenities (e.g., covered parking, interior finish, retail services), and lower lease rates. Over time, Class A buildings that are not updated can shift down into the Class B category. The opposite can also occur: Class B buildings can be upgraded to Class A status.

The key determinant for land use devoted to the office-using industry sector has been employment, particularly in finance, insurance, and real estate, and the subsector "services." Grouped together, these sectors are frequently labeled "producer" or "advanced" services. The degree to which the correlation between employment growth in these sectors and the demand for office buildings and related infrastructure (for example, parking decks) can be forecast in the future has become less predictable due to technological changes occurring in the office workplace.

First, advances in computers and telecommunications are allowing for the elimination of employees, for telecommuting for employees from home, and for electronic data storage, which decreases file storage and office space requirements. Certain sectors, notably accounting and others with employees that frequently work at clients' locations, are adopting "hoteling" or "virtual office" strategies that decrease space requirements as well. Instead of being given individual offices or workspaces, employees either share desks with others on a permanent basis or use office space that is never permanently assigned to any one employee, but, rather, is "checked out" to employees on days they are at the corporate office. It has been estimated that hoteling can reduce the amount of space required per worker from 300 or 400 square feet to less than 200 square feet.

Second, the information infrastructure required by today's advanced services and the movement toward more open office space design may mean that older office buildings are structurally unsuitable or obsolete. Also affecting the desirability and costs associated with acquiring and maintaining older structures is the 1992 Americans with Disabilities Act (ADA), which can require costly retrofitting. In addition, the desirability of property constructed before 1970, when the use of asbestos-containing material was widespread, has been adversely affected by the 1990 asbestos abatement legislation as well as EPA remediation rules (Nunnink, 1994). These factors make it more difficult to predict continuing demand for older Class B and C buildings as office use. They should alert the local economic development planner to the need for promoting alternative uses to ward off decline when they lead to weakening market interest.

The Future of Office Use in the Central Business District

The influence of advanced telecommunications and other technology on the central business district (CBD) has come under increased speculation. Traditionally, the office-using sectors were concentrated in CBDs because of the importance of face-to-face contact for highly skilled and highly paid professionals in the advanced services industries. The centralization of activity in high-rise office buildings decreased the travel distance and increased access to clients for these professionals. Some have suggested the "spatial glue" (this need for face-to-face contact) is undermined by the ability to have instantaneous communication across distance. Others have countered that tacit information exchange will continue to require face-to-face contact for "the people involved at the apex of corporate power, and also those at the apex of financial markets and professional services, [who] need to be 'in the thick of it' in ways that cannot be substituted by telematics" (Graham & Marvin, 1995, p. 142).

Two other forces have been suggested as helping to ensure the primacy of the office building stock of CBDs within metropolitan regions. First, heavy investments in valuable real estate by corporations make firms reluctant to leave CBDs because it would devalue their fixed assets (Castells, 1996). The second force is the U.S. government's Executive Order No. 12072 that, seeking to "strengthen our Nation's cities," requires federal agencies to first consider CBD locations when choosing their offices. This second force will, of course, only have a positive influence for metro areas that are the recipients of regional offices of federal government agencies, a small fraction of the nearly 300 standard metropolitan statistical areas (SMSAs) across the United States.

The first force—the notion that heavy corporate real estate investment in the CBD inhibits the tendency for corporations to leave the central business district—should not be taken for granted for at least two reasons. First, there are many well-known examples of major headquarters choosing to leave their CBD locations for suburban office parks with, perhaps, the relocation of Sears out of Chicago to the northern suburb of Oak Park being the most well-known. In an age of "lean and mean," highly expensive downtown "trophy" spaces are luxuries many firms will not maintain. They are better off to sell the real estate, even if there is some loss involved, because, on net, their costs are significantly reduced in the suburban location. In the move to achieve increased flexibility, corporations may tend toward leasing of space customized for their needs as opposed to making outright purchases.

Second, the advent of Real Estate Investment Trusts (REITS) offers even more flexibility to corporate location decisions. The multicity REIT ownership of office property releases firms from space leases that are much longer than product cycles by allowing them to transfer between office buildings owned by the same REIT. Although REITS managed only 8.3% of commercial real estate in midyear 1997, analysts predict the REIT management system will dominate real estate investments in the near future (McIntosh & Whitaker, 1998). Further, the majority of REITS are located in the suburbs, which creates another disincentive for remaining in the CBD.

While economic development planners and practitioners need to be aware of the previously mentioned trends, their true impacts will not be known for some time to come. We also do not know whether the continued development of edge cities may actually make the CBD a more desirable, cost-competitive location for office sector uses in the future. Evidence is accumulating of the overdevelopment of edge cities that, due to their suburban origins, do not have the transportation infrastructure to accommodate the growing volume of workers and shoppers. Commute-hour gridlock increasingly requires the presence of privately hired traffic police to facilitate the movement in and out of shopping centers and suburban office parks. In addition, the office buildings in edge cities increasingly resemble the trophy towers of the CBD, and the high demand for such space can result in lease rates equal to or exceeding those of the downtown that may lessen the attraction of suburban locations.

For example, the suburb of Dunwoody, Georgia, has become an edge city on the 285 freeway that circles Atlanta. Its office market, known as Central Perimeter, was larger (19.6 million square feet) in 1998 than the original downtown market of Atlanta (15.5 million square feet). Like many affluent bedroom communities that developed in the 1960s and 1970s, Dunwoody attracted retail and other service businesses that followed the population to the suburbs. This development set the stage for subsequent office sector development. Today, Dunwoody's office space quality, as indicated by lease rates, rivals or exceeds that of downtown Atlanta. Central Perimeter tenants in the advanced services sectors pay higher average lease rates than their counterparts in the downtown area. Thus, based on lower costs and easier access near points where the downtown has mass transit, the CBD may once again become a better location for a broader range of services. Alternatively, office-using service activity may simply continue its outward movement to newer and less expensive office parks being developed on the ever-expanding metropolitan fringe.

Again, while the ultimate outcome remains to be seen, it is important for the economic development planner and practitioner to be aware of these trends and to take steps to strengthen the reuse potential for the existing office building market. We return to this point later. First, however, we consider in greater detail the reasons firms may choose to stay downtown, and in particular, profiles of downtown buildings that have been retrofitted to make them suitable for today's needs.

What kinds of firms may always stay downtown? An obvious answer is those that need proximity to the courts, city hall (thus, law firms especially), and centrally located federal facilities due to the Executive Order previously described. Those seeking to draw on the technical resources and labor of central-city-located universities are another group. An area near the Georgia Institute of Technology that has been experiencing revitalization by information technology firms, many started by Tech graduates, has recently been labeled the City of Atlanta's "fiber optic alley."

Further, because the overall occupancy costs in the downtown area are now more in line with those of the suburbs, in large part due to the suburban space becoming tight enough that rents have been rising, those seeking less expensive space may stay in or return to downtown. A Chicago real estate analyst noted that despite add-on charges such as higher city property taxes, a downtown lag in rent escalation has helped keep downtown office space attractive (M. Ludgin, personal interview, July 22, 1997).

Chicago's downtown offers many examples of office building renovation and reuse. In part, this is because Chicago has a rich inventory of older (Class B) buildings with desirable amenities such as unique architectural design and historical landmark designations. Economic developers can highlight these same amenities in most older cities across the nation.

Two noteworthy examples of early office buildings that have warranted extensive renovations to meet today's demand are the Rookery Building and the Marquette Building in downtown Chicago. The Rookery Building, located on LaSalle Street, is a Class B building rehabilitated in 1992. It was originally designed by two of the early "Chicago-style" architects, Daniel H. Burnham and John Wellborn Root. When it was completed in 1888, its 12 floors made it the tallest building in the world. The Rookery Building was built by Central Safety Deposit Company, which had obtained a long-term lease (expiration date: 1982) on the site from the City of Chicago. Less than 20 years after it was built, Frank Lloyd Wright was commissioned to redo the interior and the atrium, increasing its historical significance. In 1931, William Drummond brought mechanical improvements to the building (i.e., updated elevator banks). On lease expiration, the City put the property on the market, and it was eventually acquired and renovated by Baldwin Development Company. In the late 1990s, it had a 94% occupancy rate.

Another early office building in downtown Chicago that has maintained its viability is the Marquette Building, designed by Holabird and Roche and completed in 1894. Named for Pere Marquette, the building's lobby is a memorial in honor of the Frenchman's expedition to Illinois as well as to important Indian chiefs of the Mississippi Valley. This 17-floor building was restored in 1980 by a John D. MacArthur Company. Now owned by the MacArthur Foundation, the building had a late 1990s occupancy rate of 80% and lease rates from $17 to $20 per square foot.

Not all older Class B buildings with high aesthetic appeal and good architectural features are good candidates for retrofitting. Some have floor plates that are simply too small to work for today's needs. These Class B buildings, such as one highly attractive art deco building in Chicago, are good candidates for conversion to residential use (M. Ludgin, personal interview, July 22, 1997). This is often the best strategy for the stock of Class C buildings in older cities. For example, downtown Chicago has a wealth of Class C buildings (constituting nearly 60% of the office stock), and many do not warrant retrofitting for continued office use. The vacancy rate among these downtown Class C buildings was 23% in 1997 (*Metro-Chicago Office Guide*, 1997). Because the net effective market rents for these buildings typically runs only $1 to $2 per square foot, it is not cost-effective to modernize them (Chanen, 1997). Increasingly, these Class C buildings are being converted into residential lofts and condominiums.

A recent conversion occurred in Chicago of the former corporate headquarters of the Florsheim Shoe Company (see Figure 6.1). The building is located on the edge of the prime downtown space in a commercial area. Profiles of the first 70 individuals to reserve one of the 212 units in the building indicated that it was attracting an older house-buying audience of empty nesters who no longer desired the large, single-family-home lifestyle of the suburbs (Kerch, 1997).

Examples such as this from Chicago and a late 1998 profile of Denver (see "Denver Stands Out," 1998, and Figure 5.3.) point to an emerging national trend in demand for downtown living, particularly among the aging baby-boomer cohort. Economic development planners and practitioners can capitalize on this trend to provide reuses for their localities' obsolete office buildings, as well as their growing stock of obsolete industrial properties, particularly warehouses, that we discuss in the second half of this chapter.

In fact, recent research on large U.S. cities demonstrates a reversal of postwar trends with many cities experiencing an increase in the number of people living downtown. Eugenie L. Birch's research (2001), which updates a joint study by the Brookings Institution and the Fannie Mae Foundation, shown in Table 6.1, found that between 1990 and 2000, 35 cities out of the 45 examined experienced a return to the downtown (Birch, 2001).

Demands for Smart Buildings

Office buildings do not need unique architectural features to maintain their desirability for office use if they can be retrofitted to "smart" or "intelligent" status. Increasingly, new Class A buildings are expected to be smart-intelligent buildings, and it is this expectation that raised original concerns that Class B buildings would become increasingly

Figure 6.1.   View of Conversion in Process of Florsheim Corporate
                   Headquarters to Residential Lofts

obsolete. Intelligent buildings are defined as those that have "adaptive environments of high quality, energy efficiency, security and safety, permitting optimized internal and external communications" (Gann, 1992, p. 1). The three areas of technology identified with intelligent buildings are described in Table 6.2. A decade ago, there were predictions that new technology requirements and the costs of upgrading would make much of the Class B stock obsolete. However, subsequent advances in information and building technology have made it easier and less expensive to upgrade.

There are typically 10 to 15 years of economic life for various tenant improvements (carpet, lighting, air conditioning, etc.). Thus, in each city there will be a portion of Class B buildings at any given time (because they were not all constructed at the same time) that are ready for improvements that can enhance their desirability for attracting new, and retaining old, tenants.

Recent advances are facilitating the provision of enhanced telecommunications for advanced services. Installing cabling for the enhanced

TABLE 6.1   A Rise in Downtown Living: A Deeper Look

City Name	Downtown Population 1990	Downtown Population 2000	Growth Rate 1990–2000 (%)
Albuquerque	1,197	1,738	45.20
Atlanta	19,763	24,931	26.15
Austin	10,579	10,769	1.80
Baltimore	28,597	30,067	5.14
Boise	2,933	3,093	5.46
Boston	77,253	80,903	4.72
Charlotte*	6,370	6,327	−0.68
Chattanooga	12,898	13,529	4.89
Chicago	56,048	72,843	29.97
Cincinnati	3,838	3,189	−16.91
Cleveland	7,261	9,599	32.20
Colorado Springs*	13,412	14,377	7.20
Columbus, GA*	8,476	6,412	−24.35
Columbus, OH	6,161	6,198	0.60
Dallas	11,858	15,198	28.17
Denver	4,807	6,702	39.42
Des Moines*	4,190	4,204	0.33
Detroit	34,872	35,618	2.14
Houston	7,029	7,565	7.63
Indianapolis	14,894	13,852	-7.00
Jackson	5,253	6,762	28.73
Lafayette*	2,759	3,338	20.99
Lexington*	5,212	4,894	−6.10
Los Angeles	34,655	36,630	5.70
Memphis	7,606	8,994	18.25
Mesa*	3,206	2,864	−10.67
Miami	15,143	19,927	31.59
Milwaukee	15,039	16,359	8.78
Minneapolis	36,334	30,299	−16.61
New Orleans*	6,988	8,051	15.21
New York*	153,927	170,708	10.90
New York downtown	87,743	101,493	15.67
New York midtown*	66,184	69,215	4.58
Norfolk*	2,390	2,881	20.54
Orlando*	17,965	15,999	−10.94
Philadelphia	74,686	78,349	4.90
Phoenix	6,517	5,925	-9.08
Pittsburgh	6,517	10,216	56.76
Portland	9,528	12,902	35.41
Salt Lake City	4,824	5,939	23.11
San Antonio	19,603	19,236	−1.87
San Diego*	15,417	17,894	16.07
San Francisco*	32,906	43,531	32.29
Seattle	12,292	18,983	54.43
Shreveport*	377	443	17.51
St. Louis	9,109	7,511	−17.54
Washington, DC*	26,597	27,667	4.02

SOURCE: Birch (2001). Reprinted by permission of E. L. Birch.
* Cities included in "A Rise in Downtown Living" study by The Brookings Institution and the Fannie Mae Foundation, November 1998.

TABLE 6.2   Intelligent Building Technologies

Building automation	Energy management, temperature control, humidity control, fire protection, lighting management, maintenance management, security management, access control, space planning, and management.
Office automation	Local Area Networks (LANS), electronic mail, data processing, word processing, management reporting and executive information systems, document image processing, and other internal communications such as audiovisual.
Enhanced telecommunications—connections between buildings	Digital PABX, routing cost analysis where the landlord acts as public utility for the building, teleconferencing, value-added data services (VADS), ring networks, and satellite uplinks.

SOURCE: Adapted from Gann (1992) and supplemental information provided by BellSouth, Atlanta, 1998.

telecommunications needed by the advanced services sectors has become less of a problem for older buildings than it was previously anticipated. Today, only two pairs of wires are required for digital communications: one pair for the phone line and one pair for the data line. Previously, analog phone wiring required 24 pairs of wires. Thus, the wiring is simplified, it takes up less space, and the wiring itself is also smaller, making it easier and less labor intensive to install between floors of existing office buildings (B. Schlesinger, personal interview, September 27, 1997).

The newest technology advance—wireless telecommunications—has the potential to sidestep the wiring retrofit issue altogether. At any rate, some industry experts consider wiring and information infrastructure to be less important issues in retrofitting a building than upgrading air conditioning (A/C) and vertical transportation systems (e.g., elevators). This is because the A/C systems of most Class Bs were designed to accommodate a ratio of one person per 175 square feet. Today, with team approaches in personnel organization, ratios are closer to one person per 150 square feet. This means A/C systems require upgrading (more British Thermal Units [BTUs]) because there are more bodies in the same square footage, and accompanying these bodies are more pieces of heat-generating equipment (i.e., computers, printers). However, due to changing chlorofluorocarbon (CFC) requirements and the fact that chillers wear out and need replacing, it is normal to replace A/C systems in B buildings, and this will typically not be seen as a cost-prohibitive capital expenditure for retrofitting.

As more employees are put into B buildings, demands on the vertical transportation system (e.g., elevators) also increase. Further, elevator systems are expected to work more quickly or efficiently. It is now possible to update elevator systems less expensively through overlaying rather than replacement. Overlaying involves taking out the old relay and controller technologies and replacing them with advanced solid-state microelectronics technology. This technology can increase the vertical transportation capacity of a building by 25% without requiring the installation of new elevators because it has artificial intelligence; that is, it learns the traffic pattern of a building and adjusts for it, automatically going to floors where peak traffic occurs at specific points of the day. Hence, microelectronics technology is not only affecting the organization of office work—resulting in fewer clerical personnel per professional staff—it affects the utility of buildings as well.

Upgrading the information technology of an older office building, whether in the downtown or an older suburb, increases the level of amenities it offers and helps to maintain its appeal for advanced services users. Two examples from Atlanta are illustrative. A 400,000-square-foot, 27-floor building on Peachtree Street in the downtown area was renovated cosmetically in 1998 to update its 1960s look. At the same time, its redevelopers put in smart utilities, as well as several options for access to the information superhighway. This required an upgrading of electrical service throughout the building. The renovating expense of $600,000 came on top of an asbestos abatement project for the building that cost $120,000 per floor (Silver, 1997).

The upgrading of the second Class B building in Atlanta could be seen as a major economic development achievement of the late 1990s. The office building at 55 Marietta Street in Atlanta was of typical Class B 1970s construction. It was bought out of bankruptcy by its present owners. Due to the strategic thinking of the building's new owners, it became a telecommunications hub: a vertical telecommunications industry cluster that was home to high-technology firms such as Quest, LCI, MCI, Media One, Vartec, and Avdata. The building provides technical space, not corporate image-making space. Its new owners decided to invest heavily in telecommunications network hookups.

It is on a fiber ring served by Metro Fiber, MCI, and Media One. It also has a satellite earth station on its roof. The tenants of 55 Marietta are data-intensive firms but not those associated with back-office activity.

They are secondary market telecommunications companies that take telecommunications data, bundle it, and transmit it over long-haul fiber to other cities. To them, connectivity to the outside fiber network is important, but so is the interconnection within the building. They are a part of the bill and rebill trend in telecommunications services, enjoying a symbiotic relationship with one another.

The owners' investments in 55 Marietta have paid off well: The lease rates of this Class B building have come to exceed those of some Class A buildings in Atlanta, and the building had a 1998 value of $11 million to $12 million in excess over its original purchase price. Thus, the city is a beneficiary of increased property tax revenue.

Fifty-five Marietta's success in colocating data-intensive firms stems from these firms' need to engage in "peering," whereby telecom companies have to interconnect with one another to get data from one company's customers to another company's customers. Public peering points—buildings that serve as connectors for fiber from lots of service providers—are usually very congested with traffic, yielding slow performance in data transmission. In many cases, a company will have private peering arrangements with other companies, and those companies will colocate in a building where the physical connection is made.

The vertical telecommunications firm cluster at 55 Marietta can be seen as a new form of what economic developers have traditionally called agglomeration economies. The conditions for this agglomeration economy are made possible through the installation of the right kinds of telecommunications infrastructure within the building. The economic development insight is that this appears to be a relatively easy and replicable way to attract data-intensive firms, essentially creating a vertical telecommunications office park in the central city that may be competitive with the new trend in "wired" suburban campus-style office parks.

The Reuse of Industrial Properties

The final quarter of the 20th century was characterized by the rise of an international economy that profoundly changed the organization of U.S. manufacturing production, compelling it to seek greater flexibility. Although the 1980s' office market was characterized by speculative overbuilding, increased global competition was forcing a restructuring of industry that would discourage speculative building activity. Furthermore, the nature of industrial plant construction has shifted more toward owner-specific and preleased facilities.

Although substantial offshoring of production has taken place in the last quarter of the 20th century, it is important to note there is still significant existing manufacturing activity, domestic siting of new plant locations driven—in part—by foreign inward investment, as well as shifts in the scale and scope of domestic production processes. These developments in manufacturing can affect the demand for central-city and older, inner-ring suburban industrial land use and properties. Local economic development planners and practitioners have a role to play in mitigating the negative effects and maximizing the positive effects industrial restructuring can have for older, centralized industrial property.

A major issue affecting the reuse of central-city industrial property is environmental contamination. A significant portion of the existing industrial property market is considered "brownfield," as discussed in Chapter 3. The brownfield issue influences the financial feasibility of technological upgrading of existing industrial properties. Further, remediation technologies and their costs affect the redevelopment and conversion possibilities for a given property site. Thus, the brownfield issue contributes to the unleveling of the playing field for industrial inner-city and older suburban redevelopment versus new suburban and exurban development.

Just-in-Time (JIT) Production
Versus Other Industrial Location Factors

The total industrial inventory in suburban areas is more than twice that in central cities, and this differential is expected to increase over time with continued suburbanization of industry. The movement toward flexible technologies, such as just-in-time (JIT) production systems with accompanying trends toward adjoining warehousing and distribution systems to minimize inventory costs, requires different facility designs for which existing industrial properties may not be suitable. This may be even more the case with older industrial properties found in central cities. Given historical preauto building density patterns, these properties are more likely to be bounded on all sides and unable to expand.

The prognosis for existing industrial properties appears more favorable, however, if we view the production approach's influence on requirements for building design and layout by its distinction as a "manufacture-to-order" versus "manufacture-to-stock" approach. The manufacture-to-order process is integral to the JIT plantwide production philosophy whereby production runs are set up from actual sale orders. The process entails low inventories because the product is only produced when an order has been received for it, and the product is shipped to its customer upon completion. Thus, the system requires relatively little space for inventory storage compared with manufacture-to-stock systems in which a product is stockpiled until a future order arrives for it (Myers, 1994).

Myers (1994) has observed how the difference in the two approaches translated into significantly different space requirements for a Japanese and U.S. automobile plant, each manufacturing small-to medium-sized transmission gears and having the same production output. They also had virtually the same number and type of machinery. The difference between the two plants was the Japanese implemented a manufacture-to-order production system, whereas the U.S. plant implemented a manufacture-to-stock system. However, this difference, Myers notes, meant the Ford Motor plant required 900,000 square feet for its facility, whereas the Japanese plant required just 300,000 square feet.

JIT or manufacture-to-order approaches can be, and have been, installed in existing industrial facilities of central cities that tend to have a smaller footprint, have multiple stories, or both than competing suburban space. In fact, the Allen-Bradley Corporation of Milwaukee, Wisconsin, has installed two of what it calls "factories within factories" in its original multistoried manufacturing site and headquarters (Leigh, 1996).

Many central-city industrially zoned locations have easy access to highways and greater proximity to airports than suburban locations. If overall manufacturing space requirements are shrinking through the adoption of JIT production processes, then the lower-cost vacant central-city plants could be competitive with suburban sites. In addition, if further suburban development is driving up prices and reducing availability for industrial activity, the central city and older suburb may become an attractive low-cost alternative *with all other factors being equal.* Joel Parker (personal interview, June 18, 1998) notes, however, that while the use of JIT production strategies and flexible manufacturing cells (FMC) can allow a firm to use only 25% to 50% of the space it previously took for production, the saving of space is only one consideration and unlikely to be the most important for most firms. What other factors require consideration?

On the one hand, there are considerations of public infrastructure. The public infrastructure requirements ranked highest today by type of facility suggest central cities and inner suburbs need not be ruled out as viable locations. A 1993 survey found access to an interstate highway was among the five key factors for manufacturing plants and distribution-warehouse facilities, whereas access to a major airport was among the top five for regional headquarters and sales offices (Coffee, 1994). Many central-city and older-suburban locations have easy access to highways and greater proximity to airports than fringe suburban locations. If overall manufacturing space requirements are shrinking through the adoption of JIT production processes, then these more centralized, low-cost and vacant plants could be competitive with suburban sites. Moreover, if continued suburban and exurban development is driving up prices and reducing availability for industrial activity, these centralized locations may become more attractive low-cost alternatives.

However, higher land costs, traffic congestion, quality-of-life issues, and regulatory issues can all make the central-city and older-suburb industrial market less attractive. Non-central-city locations have fewer layers of government with which to contend. Firms in central-city locations may have to contend with city and county government overlapping jurisdictions and separate water treatment districts. They may also have to contend with clean air regulations directed at the central city. According to Joel Parker (personal interview, June 18, 1998) of the International Development Research Center, "change and flexibility" are the Holy Grail to firms. They seek, above all, strategic flexibility, which means that they want to have the ability to change direction and move very quickly.

All corporate location decisions involve trade-offs. In the case of firms employing FMC technology, the number of employees required will be less, but those employed will likely need to be highly skilled. Thus, central-city and older-suburb locations may be less likely to meet the quality of life requirements for work and residential location of the high-skilled workers the firm needs to attract. In addition, today's larger firms—particularly those whose business strategies are shareholder driven—tend not to treat place as a factor at all in their facility location planning.

On the other hand, with the previously described emerging trend toward downtown residential development and efforts to make downtowns "live and work" spaces for professionals, advanced manufacturing firms may find well-developed industrial properties in more centralized locations desirable. This may particularly be the case for firms with significant R&D activity carried out by engineers and scientists, or whose chief executive officers (CEOs) prefer city living, or which are having trouble finding needed blue-collar labor in the suburbs. The development of planned manufacturing districts and office-light industrial parks around existing industrial properties described in Chapter 4 can act also as a significant force for the reuse of industrial properties.

To date, however, JIT production strategies have had their biggest impact on suburban industrial development. The industrial activity in the inner city tends to be light assembly, or light-to-heavy manufacturing. JIT may lead to a decrease in the space required for a particular production line; however, it is just as likely today's firms are engaged in efforts to provide a wider array of products (in keeping with the growth of niche markets and more diversified consumer tastes). They therefore need to run more than one production line that negates the space savings from JIT (A. Lydon, personal interview, July 17, 1997).

Thus, the promise flexible manufacturing technologies had for increasing the desirability of older manufacturing facilities is not something on which economic developers should count heavily, although there still may be some opportunities to maintain or recapture manufacturing activity to diversify the economic base. To help economic developers appreciate the challenges and potentials for industrial building reuse, we present here a discussion of general issues in plant layout.

There are two aims in plant layout for existing manufacturers:

1. To maximize manufacturing efficiency though such things as minimizing material handling and bringing workers closer together on the floor for inspection tasks

2. To incorporate new technology tasks

In either case, it does not really matter whether the firm is working with a new or older space, according to Plant Layout Specialist Bob Napora (telephone interview, July 30, 1997) with the Chicago

Manufacturing Center. The building specifications of older buildings will typically not present a problem in plant layout, unless the building dates back to the turn of the twentieth century. The buildings of this era have low ceiling heights and much compartmentalization within floors. To use these buildings, walls need to be knocked down, which is generally cost prohibitive.

If older buildings have wooden frames or weak foundations, they will not be suitable in cases in which industrial equipment needs to be installed that induces a lot of vibration. Because of the move toward the use of automated and heavy equipment, older industrial buildings may require the installation of concrete pads. The previous industrial use of the building can then become a key issue because the installation of pads requires digging down anywhere from several feet to as much as 16 feet, in the case, for example, of the installation of a steel coil fabricator. This requires obtaining a building permit and soil testing, which means potential brownfield properties are unlikely to be considered. However, "cleaner" manufacturing processes such as apparel assembly would not pose a problem for such uses.

Economic developers should be aware, however, that the adoption of new manufacturing paradigms such as continuous processing or even JIT does not always require new employees, equipment, or building renovation. It may simply entail the adoption of a different strategy that makes more efficient use of the existing industrial building, personnel, and equipment. Napora (personal interview, July 30, 1997) cited the example of a leather tannery: It had 30,000 hides in inventory for its queuing operations, or one third of $1 million in buffer. Through a "relayout" of the facility only, the company was able to reduce the number of hides in the queue to 100, thereby having only $1,000 in buffer.

Industrial Property for Warehousing and Distribution

Industrial land and building structures can provide space for production (service) activity or warehouse (distribution) activity. Warehouse space can be attached to the production facility to store inventory or can store goods for wholesaling and retailing to the market (thus, seeking a centralized location on a distribution network). The emphasis on reducing inventory, as well as the time inventory is stored in a warehouse, stems from the fact that inventory typically makes up 10% to 20% of the assets of a manufacturer, and from 15% to as much as 60% of the assets of wholesalers and retailers (Milner, 1997).

Milner (1997) offers the following rule of thumb for assigning cost in a distribution-warehouse center: Inventory carrying costs amount to 25%, warehousing costs (building and labor) are another 25%, and transportation costs (inbound and outbound) make up the remaining 50%. The emphasis on minimizing these costs while shortening the delivery time to customers made possible by new technologies and scheduling arrangements is having a significant impact on industrial property devoted to warehousing and distribution.

To elaborate, in considering transportation costs, the outbound portion has historically been the larger share because goods leaving the warehouse were typically being distributed at higher less-than-truckload (LTL) rates. This created an incentive to have more distribution centers closer to their customers to lower the aggregate number of expensive LTL miles. One of the impacts of trucking deregulation has been to significantly lower the rate differences between truckload and LTL rates. Consequently, in today's climate, a distribution center's outbound transportation costs are now minimized when the center is located at the point of minimum distance to *all* customers. Thus, for companies serving a national market, the most cost-effective strategy locates distribution centers near the country's geographic central point of population density. In recent years, major distribution centers have sprouted up in the midsection of the United States, and in particular, the cities of Memphis, Cincinnati, Indianapolis, and Louisville. Illustrative of this trend was the Pfizer company's (the pharmaceutical giant) 1995 decision to consolidate distribution centers in Chicago, Atlanta, and Dallas, with one in Memphis (Milner, 1997).

The increased use of airfreight for goods distribution is also influencing companies' location decisions. Milner (1997) notes that for high-value inventories (e.g., biomedical products, computer chips, electronics, or pharmaceuticals), the higher transportation costs associated with airfreight can be offset by the alternative high inventory carrying costs. He foresees more companies locating and relocating near major airfreight hubs to gain competitive advantages in delivery service times. Cities that will particularly benefit from this trend are those that are already major hubs: Memphis (Fedex), Louisville (UPS), and Wilmington, Ohio (Airborne).

In addition, because of greater centralization of warehousing-distribution activity, much larger warehouse space (i.e., 100,000 square

TABLE 6.3   Selected Building Attributes of Modern Warehouse
Distribution Facilities

	10 Years Ago	Today	10 Years Ahead
Clear height	24 feet	24–30 feet	24–30 feet
Sprinklers	0.33–0.45 gallons per minute	ESFR[a] —60 gallons per minute	ESFR[a] —60 gallons per minute
Dock doors/ space ratio	1.0/15,000 square feet	1.0/5,000 square feet	2.0/10,000 square feet
Truck maneuverability	115–120 feet	120–150 feet	185 feet[b]
Bay spacing	40 × 40 feet or 48 × 48 feet	48 × 48 feet or 52 × 52 feet	52 × 52 feet or 60 × 52 feet

SOURCE: Christensen and Ackerman (1996, table 5). Reprinted with permission by the authors.
a. Early suppression fast response.
b. Includes space for trailer parking.

feet or more) with automated racking systems for inventory is being demanded. The requirements of these facilities, detailed in Table 6.3, appear to make much of the older warehouse space in inner cities and older suburbs obsolete, creating a significant challenge for economic developers.

Increased use of trucking and air as opposed to rail to transport freight suggests centralized urban warehouse space located along rail spurs may not be demanded to the degree it was previously. Further, the shift to containerization in rail transport distribution makes older warehouses with smaller loading bays unusable. The trend toward larger tractor trailers, in general, makes moving through central-city streets more difficult, and older warehouse facilities may not have adequate maneuvering space to get the larger trucks in and out of the site.

The net effect of shifts in the distribution of goods just discussed is to suggest weakening demand for much of the nation's inner-city and older-suburb warehouse space. What will be the source of the diminished demand that continues for this warehouse space? There will still be demand for warehouse space to provide distribution to inner-city markets. Further, firms that do not employ the latest logistics approaches such as centralized national distribution systems, either because they do not service national markets, or their distribution strategy is less sophisticated, or other factors such as cost of space or attachment to location outweigh logistic considerations, may still provide demand for inner-city warehouse space. Firms that produce products that can be more easily transported on smaller trucks, either because the products are relatively small, are distributed in relatively smaller volumes, or both, may also continue to use inner-city warehouse space.

As Christensen and Ackerman (1996) note,

> Typically, small firms engaged in some level of distribution cannot afford the cost of developing or purchasing the best hardware and software for inventory management control. Nor do small companies need the acres of space and 30-foot clear heights found in state-of-the-art regional distribution centers. What small companies will need is well-located, no-frills warehouse/distribution space. (n. p.)

They go on to note, however, that which is considered well-located space is *suburban space.*

What is becoming of the nation's growing stock of inner-city obsolete warehouse space and the industrial land on which it sits? A positive outcome is one in which warehouses located in areas of the city undergoing commercial and residential "gentrification" are converted into retail space and loft apartments. This particularly seems to be a strong possibility for old, multistoried brick warehouses. Obsolete factory space is experiencing similar transformations. The possibilities for conversion of such space, however, can be hampered, if not stymied altogether, if the production process that occurred within the factory was one that resulted in environmental contamination.

Still, a significant portion of the nation's inner-city warehouse and factory space is not attractive enough or located well enough to warrant conversion under market conditions. Thus, cities appear to be faced with a growing problem of industrial "wasteland."

Ironically, central cities face another challenge in maintaining the viability of their industrial land when they have successful manufacturing firms. This stems from the lack of opportunity for on-site expansion. The density of building structures in central cities typically precludes possibilities to enlarge industrial facilities, or add more parking for employees, and freight-moving trucks. Acquiring nearby underused or vacant industrial property to expand creates the strong possibility that brownfield issues will be encountered that raise difficulties and the costs of expansion. This can create an impetus for successful manufacturing concerns to move to the suburbs.

Economic development planners may sometimes be able to meet the needs of firms for on-site expansion through the use of street vacation programs. Such programs focus on streets that are underused and whose removal from the street network would not significantly affect transportation problems. They are sold for a nominal amount to the firm, and the firm pays the cost of relocating utilities when necessary. The city gains twofold from street vacation programs: It retains a firm it would otherwise have lost, and it gains more tax revenue once public land is in private hands.

Black (1995b) observed that "The industrial and warehouse space markets are becoming increasingly difficult to understand and predict because of the decreasing role that employment levels play in determining space demands and the restructuring that is occurring in the business" (p. 51). We appear to be seeing a reverse of the trend in the office market: The space per employee is increasing because of improved labor productivity.

Many forces have been working against the preservation, let alone expansion, of industrial activity and full use of inner-city and older-suburb industrial property. Similar to office property, from an economic development perspective, there is genuine validity to the notion that a healthy local economy requires well-used industrial property. Such property holds the economic activity that helps to diversify the local economy. It provides industrial jobs that often pay higher wages to the lower-skilled worker than does the services sector. Maintaining a strong industrial base in the inner city and older suburbs helps to stem the development of more brownfields. Further, reusing industrial property for another industrial use can help to hold down the remediation costs of industrial brownfields because the use of risk-based cleanup standards allows for lower levels of cleanup when the property is not going to be used for a purpose that involves intense human contact such as housing and playgrounds.

Boxes 6.1 and 6.2 illustrate two reuse examples of very large industrial properties, one initiated by a private company and the other initiated by a nonprofit corporation. The second example illustrates a parallel opportunity to that of the 55 Marietta office building renovation to facilitate the colocation of similar firms, thereby encouraging development of an industry cluster.

Box 6.1
Profile of a Chicago Manufacturer's Within City Relocation

Tripp Lite, also known as Tripp Manufacturing, is the second largest producer of power protection products in the world and largest producer of surge protectors. (American Power Company is number one.) It is a privately owned company with annual sales exceeding $100 million. It owns stock in all of its publicly held competitors. The company operates on a model of flexible manufacturing and complete outsourcing of its products. Its 330 employees are engaged in sales, warehousing, administration, and R&D. They do the design work, quality control, and marketing; assembly is done primarily with subcontractors around the metro area who employ up to another 1,500 employees.

Tripp Lite needed to vacate its North River location in Sexton lofts (around 350,000 square feet) which it was leasing. With search assistance from the City of Chicago's Industrial Planning and Development Office, Trippe-Lite decided to purchase and relocate to the vacant Spiegel Warehouse and Administrative facility in the Bridgeport area of the city. At the time of purchase, the Spiegel facility had close to 2 million square feet and consisted of 15 buildings on both sides of 35th Street. Relocation plans called for seven buildings with 750,000 square feet to be torn down, eight other buildings encompassing 821,000 square feet to be gutted and redeveloped, and another 456,000 square feet to remain untouched. The main building on the south side of 35th Street houses corporate offices, research and laboratory facilities, and warehousing. The first floor of the building is used for shipping (it has 22 loading docks), the second and third floors are warehouse space, the fourth floor is devoted to overstock, and the fifth floor will hold corporate offices. Floors 1 through 4 have 12-foot ceilings, while Floor 5 has 20-foot ceilings, and its height affords good views of the east, west, and north Chicago skylines. Floors 6 through 11, left un-redeveloped, may be sublet to other tenants in the future.

Because Tripp's main building was previously a warehouse, it had very limited communications facilities. Ameritech ran fiber-optic cable under the ground to the building while Tripp paid to run it up through the floors of the building. The cost of doing so, according to Tripp's Administrative Manager, Bob Mazalin, of approximately $300,000 to run cable for data and voice–phone lines for 1 million square feet, was reasonable. In their previous space, it had cost $45,000 to run cable for 10,000 square feet. A big difference in the cost is attributable to the fact that it is warehouse, or relatively unfinished space, through which the cable will be run.

Tripp needed to find new space because it had outgrown its North River location. Tripp's owner lives in the city and wanted to stay in the city. Further, the company's employees primarily live in the city. Some of the buildings torn down on the new site were for the purposes of creating adequate parking for Tripp's employees. A major consideration of Tripp's owner was that the company move to a large enough space that it would accommodate all future expansion needs. Even with the demolition of buildings and the possible sale of the two former administrative buildings on the north side of 35th Street (they appear to be possible loft conversions as there are already loft developments a little farther along the street), it appears very unlikely that the Spiegel site would ever be outgrown. (B. Mazalin, personal interview, July 30, 1997; Podmolik, 1997)

Lessons for Economic Development

As we stated at the beginning of this chapter, local economic development planners and practitioners have a strong stake in facilitating the reuse of their communities' office and industrial properties. To do so, they need a proactive approach for encouraging property reuse, thereby stemming high vacancy rates among older industrial and office properties and the abandonment of properties that fuels economic decline.

Inner-city vacancy rates in both industrial and office properties exceed those of the suburbs. Continued suburbanization poses a persistent challenge to inner-city and older-suburb economic developers trying

Figure 6.2.   View of Two Buildings at Tripp Lite's New Location: Spiegel Warehouse (Back Building) and Former Spiegel Administrative Offices (Front Building)

Figure 6.3.   View of Some of the Demolition Activity Tripp Lite Undertook Around the Main Warehouse

---

Box 6.2
From a 19th-Century Rope Factory to the
Greenpoint Manufacturing and Design Center

At 1155 Manhattan Avenue in New York City, a 19th-century rope factory that eventually grew to encompass eight buildings has been redeveloped by a nonprofit organization into the Greenpoint Manufacturing and Design Center. It provides affordable workspace to woodworkers and artists. The center is located in the Greenpoint neighborhood, "a blue-collar neighborhood with a population that is largely Polish, Irish and Hispanic" ("Off the Urban Rust Heap," 1999, p. 8).

The complex of buildings became the city's property in 1972 due to tax delinquency of the then-current owners, Grossen Dye Works. The City eventually closed down portions of the complex for code violations. City estimates for the cost of the needed renovations were $14 million. The City spent $750,000 for environmental cleanup. When the nonprofit corporation formed to take over the building complex in 1994, only 40,000 square feet were usable. The nonprofit corporation acquired the complex for $1, and the City set aside an additional $1 million to pay for needed safety improvements. The center's renovation has ended up costing half of what the City estimated it would require. The Center's 360,000 square feet of space are now leased out. A total of 460 people are employed by the center's tenants, which amounts to around $800,000 in city and state payroll taxes. In addition, the center contributes $135,000 a year in city property taxes.

The Greenpoint Manufacturing and Design Center provides an example for economic developers of how old factory complexes in central cities or older, inner-ring suburbs can provide affordable space to small manufacturers and craftworkers. ("Off the Urban Rust Heap," 1999)

---

to maintain and strengthen their local economies. A substantial portion of this chapter has been devoted to identifying the space use trends and retrofitting considerations planners must be familiar with to identify potential trends of decline and to facilitate successful reuse of properties to ward off or reverse decline.

The emerging trend in downtown living, particularly among the aging baby-boomer cohort, is a trend economic developers can capitalize on to provide reuses for their localities' obsolete office buildings, as well as their growing stock of obsolete industrial properties (and warehouses, in particular).

Upgrading the information technology of an older office building, whether in downtown or an older suburb, increases the level of amenities it offers while lowering the cost of maintenance, thereby helping to maintain its appeal for advanced services users. Recent technology advances make it easier to provide upgrading. Further, as Chicago shows, even the very oldest office buildings can be upgraded, and the desire to do so derives from quality architectural design that has stood the test of time. The buildings of the 1960s and 1970s may have the lowest chances of being upgraded because they were typically designed in relatively featureless, modern box styles, and their ceiling heights are lower than those of buildings in the eras before or after. This should serve as an important lesson not only for the reuse of old buildings, but the design of new ones. Further, it illustrates how important good urban design and architecture are to sustaining strong local economic development patterns.

Demand for older manufacturing facilities may still be found among smaller manufacturers, particularly those serving inner-city markets, those making smaller products that are easier to transport in smaller trucks used to navigate city streets, or both. Reuse of such facilities is complicated by the high probability some form of contamination will be encountered (as we noted in Chapter 3). However, economic developers can assist in resolving these issues, as Chapter 4 strategies detail. In some cases, they can also facilitate the needs of firms to expand on-site through programs such as street vacations.

The net effect of shifts in the distribution of goods discussed previously suggests there will be weak demand for much of the nation's inner-city and older-suburb warehouse space. There should still be demand for warehouse space to provide distribution to inner-city markets, and for firms that do not employ the latest logistics. Further, older, centralized warehouse space may be required by firms using rail, small truck transport, or both for their goods. However, a significant task for economic development in the future will be encouraging alternative uses to warehousing for this space.

In the end, economic developers may realize their greatest successes through encouraging the reuse of properties by industry clusters, whether secondary market telecommunications firms in Atlanta or woodworking firms in New York City. In keeping with the definition of economic development planning that guides this book, however, we conclude this chapter by identifying what may well be the biggest challenge for planners and practitioners seeking to promote reuse of office and industrial properties in inner cities and older suburbs. That is, the lion's share of demand for reuse of office and industrial properties will come from the private sector. Successful private developers are profit maximizers and operate on the premise that much of a potential reuse project's final value will be determined by its location. Thus, properties located in a gentrified area, or area with high potential to gentrify in the near future, will be those focused on by private developers. A guiding principle of the definition of economic development in this book, however, is economic development seeks to lessen inequalities. Unless efforts are made to promote the reuse of industrial and office properties in the most marginal or least likely to gentrify areas, the very act of redeveloping and reusing such properties in their strongest areas will serve to create greater inequality within older cities and suburbs. To counter this possibility, local economic development planners and practitioners can seek opportunities to invest public development dollars in the reuse of buildings in the areas the private sector will avoid, essentially following a local version of the federal policy to site federal buildings in the central cities. They can also promote the use of public incentives to motivate the private sector to develop in areas that otherwise would be unlikely to experience redevelopment.

Appendix

Issues in the Retrofitting of Office Buildings

Jonathan Hoffman and Nancey Green Leigh

The private or public developer faces many options and issues when retrofitting an office building for today's business requirements. Modernizing cabling, elevator group supervisory systems, HVAC, and lighting promise significant cost savings and efficiency gains by taking advantage of the latest building systems technology. Economic development planners and practitioners seeking to promote inner-city and older-suburb revitalization will need to be just as familiar with retrofitting issues as they need to be of building codes and zoning ordinances.

Integrated Cabling

In modernization plans, the cabling associated with separate components of business systems (voice, data, video), building management systems (HVAC controls, lighting, security, elevator control), and fire and life safety are frequently combined for cost savings. The combined cabling, through conduit, cable tray, raceway, or other method, allows quicker installation at the time of retrofit and avoidance of recabling with each new tenant.

The cabling can run beneath the floor through underfloor power grids or raised access floors, through furniture, or more rarely, in the ceiling. Installing raised access floors in building retrofits has proved problematic for some older buildings that were not built with extra area between floor slabs. This is particularly the case for 1960s and 1970s office buildings that were built with lower ceilings than their predecessors and successors. The 1980s-era buildings returned to more historic standards of 14 feet between floor slabs, allowing two feet floors and ceiling while still maintaining 10-foot ceilings. Recently, however, even retrofitting buildings with little to no extra area between floor slabs has become more feasible as manufacturers have begun to offer raised access floors with only a few inches of clearance.

Structured cabling systems can reduce initial retrofit construction costs in an intelligent building by up to 30%. Most savings, however, stem from the shorter installation time that results in allowing space to be occupied earlier. These savings are dependent on the acceptance of an internationally standardized structured cabling system.

Regulations in some areas may prohibit the integration of fire and life safety systems with other building systems. The U.S. 1996 National Electrical Code allows in many cases fire and life safety systems sharing cabling or raceways with other systems. In many areas, however, local law prohibits this practice or requires special permits. Substantial savings are still possible through the integration of the remaining systems (Reinbach, 1993b).

Elevator Group Supervisory Systems

Inefficient elevator supervisory systems are frequently unable to accommodate the demands of increased population density in office buildings, especially in buildings with few but large tenants. Current rules of thumb for elevator service include the ability to move one sixth of the building's population in 5 minutes during the morning peak period, and a maximum waiting time of 30 seconds. Modern supervisory systems can efficiently operate existing elevators to handle modern traffic needs while avoiding expensive elevator replacement ("Modern Traffic Handling Capability," 1986).

High-Efficiency HVAC

Heating, cooling, ventilation, and air-conditioning systems (HVAC) have a major impact on retrofit costs. In initial construction, modern HVAC equipment purchase and installation costs can account for nearly one half of a retrofit budget. Continuing operation is even more costly. Many pre-1970s office buildings were not designed to handle internal heat sources like computers and copy machines. In addition, energy systems assumed continued unlimited and cheap energy sources. These older buildings consume 28 to 32 kWh per square foot/ year, high above the 1980s-era building average of 18kWh per square foot/year.

High-efficiency HVAC systems include variable air volume, variable speed drives on supply and return fans, and a division of a building into many different heating zones. Reflective solar control film on exterior glass frequently complements these systems. Combined with other energy-saving tactics, inefficient buildings that are retrofitted can save as much as 72% of their energy costs.

Realizing these savings may not come cheap, however. HVAC modernization can cost $500,000 to $1 million, and for some buildings cost-benefit analysis may indicate savings are not large enough to warrant retrofitting. Yet, in such cases, some local utility companies find it cheaper to subsidize energy retrofits than to build a new power station. For example, in 1991, the Los Angeles Department of Water and Power paid 49% of a $983,000 energy retrofit of an inefficient 110,000-square-foot office building (Reinbach, 1993a).

Lighting

Modern lighting retrofit involves more than a switch from incandescent to fluorescent lighting. Typically, older 40-watt, T-12 lamps are replaced with more efficient 32-watt, T-8 lamps, and electromagnetic ballasts are replaced with electronic ones. These lamps are combined with optically designed specular reflectors and deep-cell aluminum parabolic louvers.

Installing electronic occupancy sensors can also cut lighting costs up to 30%. These sensors, which automatically switch off the lights when there is no movement detected in rooms, are particularly effective in those rooms with varied and unpredictable uses like private offices, conference rooms, and rest rooms.

It is important to note, however, that the operating savings available from a lighting retrofit will depend primarily on the billing practices of the local utility. Most utility companies add on customer and demand charges, sometimes up to 60% of the total bill, which are not tied to consumption levels. Time-of-day and block rates may also mean that a 25% drop in energy consumption will not equal a 25% cut in the energy bill (Mumford, 1993).

7

Job-Centered Economic Development: An Approach for Linking Workforce and Local Economic Development

Workforce development is more than job training. It refers to the constellation of activities required to place people who have few job skills, or skills that have become outdated, into gainful employment. Among these activities are orientation to the work world, skills training, placement, mentoring, and postemployment services. Workforce development links to the principles of raising standards of living and reducing income inequality presented in Chapter 1. In principle, few would argue with the need to link economic development and workforce development. A skilled workforce makes every planner's list of elements of a "good business climate." Yet for the majority of municipal governments, workforce development is functionally separate from economic development.

At some level, it makes sense for the two to be separate. Economic development is place based and typically consists of land use planning, economic analysis, and working with private sector employers and developers. Workforce development focuses on individuals and requires coordination with social service agencies and other public sector organizations. Another reason for functional separation is the organizations and institutions of education and employment training—high schools, community colleges, proprietary schools, unions, and community-based organizations—are not under the influence of planning departments, or even municipal governments. In fact, the goals of education and training providers and those of economic development agencies may not even be perceived as mutually complementary (Fitzgerald, 1999). Professionals in each of these fields view their work responsibilities quite differently. Economic development practitioners typically work with developers or businesspeople. Job-training providers largely serve poor people, offering a second chance for those who failed in or were failed by the education system.

Our experience and research reveals that economic development practitioners typically do not welcome what they perceive to be social work tasks added to their jobs. For example, Fitzgerald and Patton (1994) documented a case in which staff of an economic development agency sabotaged a set-aside program created within their industrial training program that targeted young African American men. The economic development planners previously worked exclusively with employers in establishing customized training programs. Their professional identity, in terms of establishing relationships, was with businesses. They resented this new intrusion into their work as they saw job training as being under the rubric of social work, not economic development.

Professional identification is only one aspect of the functional separation. Organizational identification is another. Even if a need for linking economic and workforce development is agreed on, collaboration requires cooperation among departments within the same governmental unit and also among organizations with very different cultures and measures of success. Funding sources for workforce and economic development are different, and each funding stream has different reporting and evaluation requirements. Social service agencies measure their outcomes by services provided to individuals. Education institutions measure results through graduation rates. Training providers typically gauge their success by job placement rates. Economic development agencies may keep track of jobs created or maintained. Linking all of these functions would create a difficult task of coordination and evaluating outcomes. Nevertheless, many advocates for the poor argue that workforce development programs need better links to economic development if they are to do more than prepare people for low-wage jobs with little advancement potential.

Giloth (1998) use the term *job-centered economic development* to define a strategy for connecting economic and workforce development to achieve social justice goals. This approach focuses on identifying and accessing good jobs, building career ladders, networking, managing cases, enabling job retention and advancement, promoting innovation among employers, and advocating policies in support of living-wage jobs. Job-centered economic development asks what a workforce development system would look like if it were designed to move people out of poverty, as opposed to simply moving people off welfare. This contrasts with more traditional approaches to workforce development that focus on the needs of employers, with little concern about job quality and possibilities for moving people out of poverty (see Fitzgerald, 1999).

Because most of these initiatives are relatively new, we are only beginning to learn how job-centered economic development works. In this chapter, we use a set of principles that underpin best practice in workforce development (see Fitzgerald, 1999) to analyze a job-centered economic development initiative in Seattle, Washington, and a related initiative in suburban King County (the Seattle and King County Jobs Initiatives [KCJI]). Much can be learned about the viability of the job-centered approach from these initiatives. To the extent that workforce development focuses on the disadvantaged, it is implicitly equity focused. However, it is only in addressing structural economic issues that workforce development can work toward the goals of raising wages and reducing income inequality. The Seattle and KCJI cases allow us to examine the extent to which systemic change can be achieved under existing legislation, and how practitioners can shape policy to achieve equity goals. Much of what practitioners are able to do is fixed by existing legislation. These cases provide insights not only on how practice can shape legislation to achieve greater equity, but also on how to work within existing legislation to achieve greater equity.

The Legislative Environment

Workforce development is framed by federal legislation, although there is some latitude at the state level for adjustments to suit local labor market needs. Two major pieces of legislation will change the landscape of workforce development well into this century: the Workforce Investment Act of 1998 and the Personal Responsibility and Work Opportunity Reconciliation Act of 1996.

It has been acknowledged since the 1980s that the federal job-training system is fragmented and inefficient (see Fizgerald & McGregor, 1993, for a review of this literature). In 1995, the General Accounting Office identified more than 160 federal job-training programs run by 15 separate government agencies. After passing several less comprehensive workforce development initiatives, Congress enacted the Workforce Investment Act (WIA) of 1998 to consolidate federal job training, adult education, literacy, and vocational rehabilitation programs into a more streamlined and flexible workforce development system. Further, WIA is a shift from a focus on providing a second chance for the poor or displaced workers to achieving the economic development goal of strengthening regional economies. State Workforce Boards, which equally comprise business representatives and representatives from education, social service, and labor and community organizations, work with governors in developing plans for development and continuous improvement of state workforce investment systems. Local Workforce Investment Boards (WIBs), with similar compositions, plan and oversee local workforce systems. Local WIBs cover similar geographic areas as the service delivery areas under JTPA, including all urban, suburban, and rural labor markets in a state. To receive funds, states submit 5-year plans for continuous improvement of statewide workforce investment systems.

At the core of this legislation is the One-Stop Employment Center. One-Stops, as they are called, are the centralized point of access for all federally funded employment programs. The overriding goal of the One-Stops is to place people in jobs as quickly as possible. Individuals use One-Stops to find jobs and to access occupational education programs and career development services.

A key component of One-Stops is computerized information systems through which employers post job openings. Unlike previous programs that focused on the poor, One-Stops provide universal access, making them the point of entry for all job seekers. The majority of clients are eligible for only a core set of job-search services. Those needing various social support services are referred to other agencies, through which they must separately establish eligibility. Limited job training is available for those unable to find employment. Clients who do not find jobs after receiving universal services are eligible for intensive services such as career counseling, job clubs, and résumé assistance preparation. If intensive services do not produce a job, those clients who meet income and other eligibility criteria are given training vouchers (called individual training accounts) to pay for training by approved providers.

Critics point out that an assumption guiding the One-Stops—that the major barrier to employment is lack of information about job opportunities—is wrong. Indeed, considerable research suggests that people find jobs more through social networks than classified ads or other job listings (Mier & Giloth, 1983; Granovetter, 1985; Neckerman & Kirschenman, 1991; Moss & Tilly, 1996; Holzer, 1996; Peck, 1996; Harrison & Weiss, 1998). The critics suggest that a computer-based system cannot replicate the personal interaction and access to employers through which the majority of jobs are found, particularly for poor people with weak social networks.

Another criticism is the One-Stops do not provide sufficient case management to ensure clients are receiving the best mix of services, or even the most essential services. Previous experience with voucher-based programs suggests they provide no structure for monitoring and evaluating a client's progress, either during or after training (Osterman, 1988). Further, many clients left to their own devices in finding jobs and training have few skills to navigate the computer-based systems or to make informed choices on training options.

Workforce development is also influenced by the Personal Responsibility and Work Opportunity Reconciliation Act of 1996, which replaces Aid to Families with Dependent Children with block grants to states to provide Temporary Assistance to Needy Families (TANF). The legislation limits the total time an individual can receive TANF income support to 5 years[1] and requires states to move a rising proportion of their caseloads each year into approved work activities to meet this goal. The policy of quickly engaging or re-engaging the unemployed into the world of work is known as "work first." Rather than providing lengthy job training before employment, the policy assumes that clients will move into better jobs as they gain work experience. In 1997, states were required to have 25% of single-parent families and 50% of two-parent families in work-related activity. The percentages increase to 50% and 90% by 2002. Work-related activity includes subsidized and unsubsidized employment, or employment training, but it must also include working 30 hours per week.

The goal of both the Workforce Investment Act and the welfare reform legislation is to move people into jobs quickly, with little emphasis on the quality or wages of the jobs. Critics of the "work first" orientation suggest that without education, training, and social support services, only a small percentage of TANF recipients will achieve sustained employment, let alone advance into higher-paying jobs (see Hershey & Pavetti, 1997; Pavetti, 1997; Fitzgerald, Putterman, & Theodore, 2000). Thus, it seems the existing legislation that funds the bulk of workforce development is not conducive to the goals of job-centered economic development. We turn to the case studies to examine the extent to which this goal can be achieved within the constraints of existing legislation. Further, we examine how practitioners can use the flexibility allowed by states in developing workforce investment plans to incorporate this goal.

The Seattle Jobs Initiative

Seattle is one of six cities[2] participating in the Annie E. Casey Foundation's 8-year Jobs Initiative. Started in 1995, the $30 million Jobs Initiative supports local government, community organizations, and educational institutions in pursuing a systems reform agenda to promote the creation of training programs for jobs with advancement potential that provide family-supporting benefits. Although there is considerable debate over what constitutes a living wage, the realistic goal of starting relatively low-skilled individuals at $8.00 per hour was set as the standard for the sites. Each site receives approximately $700,000 per year for planning and implementation and must provide matching funds from other sources. The funding includes an 18-month planning process, a 3-year capacity building phase, and 4 years of implementation.

Unlike many cities, workforce development was already a central component of the mission of Seattle's Office of Economic Development (OED), which administers the Seattle Jobs Initiative (SJI). Formed in 1993 under Mayor Norman Rice, OED's mission was and still is the following:

> To use the power of City government to support a healthy diversified economic base and to bring economic opportunities to all Seattle citizens—especially the most disadvantaged. To take these actions in partnership with private sector firms, community-based organizations, and other public sector institutions wherever possible. (City of Seattle, Office of Economic Development, 1994, n. p.)

Three elements are involved in achieving this mission: strengthening the economic base and business climate (business retention and attraction), supporting employment opportunities (education and job training), and supporting community-based economic development (neighborhood revitalization).

The commitment to integrating workforce and economic development began with former Mayor Rice. A former Urban League job developer, Rice knew urban labor markets did not work for low-income communities and reorganized his department of Economic Development to achieve this end. Mary Jean Ryan (personal interview, August 1998), director of OED, further explains that she and her colleagues were frustrated that their department provided public funds for job creation but had little control over who got the jobs. Part of the problem, as in many cities, was that programs targeting low-income populations were housed in the Human Services department. Thus, even employment-related concerns of poor people were categorized as social work rather than economic development.

To gain more control over job-training funds, Ryan transferred control of $500,000 of Community Development Block Grant and general funds for employment services for low-income populations from the City's Department of Housing and Human Services to OED. This decision was controversial as the perception of Human Services staff was that they cared about people whereas OED staff cared only about business. Ryan demanded more accountability in how the funds were being spent, introduced performance-based contracts for employment training, and focused performance on job placement and retention. This was a first step in linking economic and workforce development, but Ryan was searching for a way to make more systemic connections between employers and people in need of jobs. As the central intermediary for the SJI, OED would be given a chance to make these connections.

Program Overview

The SJI is attempting to link low-income Seattle residents with living-wage jobs and to improve employment and training systems to produce better results for low-income job seekers. Four goals were established in the application proposal:

1. To help residents obtain jobs that pay living wages of at least $8 per hour plus benefits
2. To ensure long-term job retention

3. To secure employer involvement to ensure that people receive relevant training, that jobs exist at the conclusion of training, and that employers get skilled workers

4. To integrate human services with employment and training services

A 20-member Leadership Committee developed an implementation plan for the SJI. Member organizations included government agencies, community-based organizations, community residents, higher education institutions, elected officials, labor, and business. The Leadership Committee identified an impact community of six contiguous distressed neighborhoods for initial implementation. The population of the community is 32% White, 30% Asian Pacific Islander, 29% African American, 7% Hispanic, and 2% Native American. Within the community, income eligibility cutoffs were defined by family income at 52% of the State Median Income, adjusted for family size. As required by the Annie E. Casey Foundation, services are targeted to people between the ages of 18 and 35[3] with low levels of education, limited employment histories, or both.

During the planning phase, which lasted from November 1995 through March 1997, SJI developed its governance structure, a 15-member executive committee divided into three subcommittees. The subcommittees analyzed the regional economy, pinpointed employment barriers of the target population, and identified programs to connect the target population to employment opportunities. Civic groups, educators, community organizations, business representatives, and government officials are involved in governance. During the 3-year capacity-building phase, beginning in April 1997, programs were implemented, evaluated, and adapted. During the final implementation phase, programs are continuing and SJI is advocating for systems change and policy reform.

Two key strategies were selected in the planning process. The *employment linkage strategy* focuses on immediate job placement for work-ready residents. This strategy leverages city investments and other economic development initiatives to increase opportunities for low-income job seekers. SJI staff identify employers willing to provide job opportunities for participants, particularly through first-source hiring agreements.[4] The *targeted sector strategy* focuses on how to structure job training, support services, and case management to place residents in jobs in growing sectors or occupations with well-paying jobs and opportunities for advancement. Other strategies existed in the early phases, but these two became the primary focus of the SJI. The next section uses the principles of best practice to analyze the specific components of the SJI.

Applying the Principles in Seattle

There are many case studies of best practice in workforce development. Unfortunately, trying to replicate a program from another city or state is difficult. Best practice is the result of more than program elements; it requires good working relationships among participating organizations. Case studies seldom present the unique history of relationships between and within organizations and agencies involved in the program. The principles were developed to provide a framework for analyzing workforce development systems rather than presenting a single best practice model for replication. Systems can look quite different and still encompass the same principles. The principles allow comparison of how different cities are achieving the same goals.

Effective workforce development programs target industries or occupations that offer competitive wages and opportunities for advancement. Sectoral strategies identify industries or occupations that have growth potential, provide well-paying jobs, and develop integrated strategies for expanding them in a state, regional, or local economy (see Chapter 2). The sectoral strategy is at the core of SJI's approach. Based on industry growth rates, number of entry-level job openings, wage rates, and possibilities for advancement, four targeted sections were originally identified by SJI: construction, manufacturing (general manufacturing, electronics assembly, computerized numerical control [CNC] machine operation), business-office occupations, and health care (certified nursing assistant, medical-clerical). SJI is the intermediary that organizes employers, community colleges, community organizations, and residents around the workforce development needs of each sector. The tasks involved are analyzing demand, determining workforce development needs, and either identifying existing training programs to connect with or creating new ones.

Sectors are added and eliminated as demand dictates.[5] The construction strategy has remained solid since SJI started, with approximately 5,340 openings per year through 2001, and 4,125 per year predicted from 2001 through 2006. SJI places clients in existing preapprenticeship programs offered by PortJOBS, the Apprenticeship Opportunities Project, and tries to increase the percentage of preapprenticeship graduates who enter into union-sponsored apprenticeships. During the capacity-building phase, SJI, PortJOBS, and the King County Labor Council (the local AFL-CIO affiliate) assessed the apprenticeship system and made recommendations for improving it.

The health care strategy was phased out because of declining employment opportunities due to mergers, cost cutting, and lack of advancement opportunities. New initiatives are continually being explored and started. A program started in 2001, Tech Talent (discussed later), is training participants for high-demand and well-paying web design positions. A program to advance hotel housekeepers to banquet servers is also under way.

Effective initiatives involve a network of providers that offer the comprehensive range of services needed to help the unemployed and underemployed secure jobs. In *Building Bridges,* Harrison, Weiss, and Gant (1998) examined how community-based training providers network with employers and other organizations to provide a comprehensive package of services that connects low-skill and low-income people to employment. They present nine case studies that illustrate different strategies and approaches community-based training providers take in forming workforce development networks. This and other research reveals that job training is not the only, or even the most important, intervention needed to connect low-income populations to jobs. People with little or no labor market experience need to master "soft"[6] employment skills (Pouncy & Mincy, 1995; Moss & Tilly, 1996). In addition, a range of social support services (e.g., drug treatment, psychological counseling) are sometimes needed before long-term employment can be an option (Fitzgerald & Rasheed, 1998). Child care and transportation are also key needs (see Osterman, 1993; Pavetti & Acs, 1996; Strawn & Martinson, 1998).[7] Postplacement support is essential to keep clients on the job and to identify opportunities for advancement. Case managers are needed to coordinate all these services. The case manager contracts with service providers, keeps in contact with the client before and after job placement, resolves problems, and works with network providers to ensure that their services meet client needs.

Few CBOs or education institutions are equipped to provide all these services. The most effective providers create partnerships to piece together a package of services to connect the unemployed and underemployed to jobs. A key difference between a networking and a "One-Stop" approach is one organization takes on the responsibility of facilitating the participation of the network partners. Potential partners and their roles are identified in Table 7.1. The partnerships created in any given city are a function of the historical relationships established among the different organizations.

The SJI, as part of the OED, is the focal intermediary that coordinates the network of service providers in Seattle. In other Jobs Initiative cities, different organizations play this role, including a CBO, a state government agency, and a charitable foundation. The role each organization plays also varies. In Seattle, almost all education and training is provided by community colleges and vocational institutes. In contrast, CBOs in Chicago, New York, Newark, and other cities have a long tradition of providing training in addition to support services (Fitzgerald & Sutton, 2000). This practice never emerged in Seattle, so clear lines of responsibility between community organizations and community colleges have been easy to maintain.

The SJI partners and their roles are listed in Table 7.2. The CBOs provide recruitment, assessment, case management, job placement, and job retention services. They also provide pre-employment (soft skills) training before participants enroll in education or training programs and employ job developers to place job-ready participants. The CBO partners think of themselves as "virtual one-stops," because they provide a full range of services to move clients into sustained employment.

Because there are many industries with strong unions in Seattle, labor has a stronger role in the SJI than in workforce development initiatives in most cities. The King County Labor Council (KCLC), which represents approximately 150 AFL-CIO locals, began with a largely advisory role in the SJI. KCLC ensures that SJI honors the terms of existing collective bargaining agreements and does not displace workers covered by such agreements. Likewise, the SJI agrees to invest in worker skills rather than creating contingent unskilled labor to compete with union workers. Recently, the Worker Center, the economic development arm of the KCLC, has taken a lead role in developing an upgrade initiative for unionized hotel workers.

Although the OED is the focal intermediary around which the SJI is organized, several of the partners also play intermediary roles in the

TABLE 7.1 Potential Partners in Workforce Development Systems

Partner Organization	*Roles*
Community colleges	Offer certificate and associate degree programs Provide career counseling and job placement assistance Provide short-term customized training for incumbent workers Provide technical assistance to employers
Community-based organizations	Recruit community residents for training programs Provide basic literacy tied to education and training programs Offer career counseling and assessment Refer residents to support services while in education or training programs Provide job training and placement assistance Provide job-keeping skills programs for clients after placement
Social service agencies	Provide transportation Recruit community residents Refer for health care (including mental health) Provide day care
Economic development agencies and organizations	Identify emerging employment and training needs of local employers Identify key industries and occupations for building comprehensive economic and workforce development programs Recruit employers to advise on education and training programs
Employers	Advise on curriculum Encourage students through job shadowing and mentoring programs Provide internships for students and teachers Establish hiring agreements
Labor	Advise on curriculum Work with employers on restructuring occupations Establish new points of entry for apprenticeship programs
Universities	Offer baccalaureate programs for graduates of associate degree programs in technical fields as part of 2+2+2 tech prep programs Serve as intermediaries in developing integrated workforce development systems Provide data analysis on workforce related issues
High schools (school-to-work or career programs)	Provide technical preparation (tech prep) curricula Encourage students to pursue careers in technical fields through career awareness, internships, etc. Provide college and job placement assistance

SOURCE: Adapted from Fitzgerald and Jenkins (1997).

sector strategies. Intermediaries are essential to the success of a networking strategy, and their role is worth discussing in detail.

TABLE 7.2   Partners in the Seattle Jobs Initiative

Partner Organization	Roles
Community and Technical Colleges Community and Technical Colleges of Seattle-King County (North Seattle, South Seattle, Seattle Vocational Institute), Shoreline Community College	Offer technical skills programs (mostly 12–16 weeks) in the targeted occupations (with credit toward certificates or degrees in some cases) Provide basic skills and ESL courses
Community-Based Organizations Central Area Motivation Program, Asian Counseling and Referral Services, Wash. Coalition for Citizens w/Disabilities, YWCA, Center for Career Alternatives, Downtown Human Services Council, Seattle Indian Center, First Place School, TRAC Associates, Refugee Federation Service Center, United Indians of All Tribes	Recruit community residents for employment programs Provide skills assessment Offer career advisement counseling Provide job readiness training Provide support services Provide case management Provide job placement assistance Provide postplacement services
Social Service Agencies Dept. of Social and Health Services (Seattle-King County), Seattle Human Services Dept., Seattle Human Services Coalition, Child Care Resources, Tender Loving Care Low Income Housing Institute	Administer the Washington WorkFirst program Provide child care information and referral Arrange sick child care Provide housing
Economic Development Agencies and Organizations Seattle Office of Economic Development Port of Seattle's Office of PortJOBS	Identify employment and training needs of local employers Identify key industries and occupations for building new workforce development programs • Recruit employers to advise on training programs • Create linkages to City's economic development programs • Organize employers around workforce and economic development needs

(Continued)

Table 7.2 Continued

Partner Organization	Roles
Employers Washington Aerospace Alliance, American Electronics Assn., individual employers	Advise on training curriculum Encourage students through job shadowing and mentoring programs Provide internships for students and teachers Establish hiring agreements Develop partnerships in union-represented industries
Labor King County Labor Council (AFL-CIO)	Identify skill and training requirements in entry-level union jobs in identified sectors Advocate for higher wages and benefits
Universities University of Washington	Provide labor market analysis for choosing target industries Provide program evaluation
Related Government Organizations Dept. of Health and Human Services, State Department of Employment Security (Puget Sound Region), State Department of Community, Trade and Economic Development, King County and Seattle Housing Authority	Provide bus tickets, gas vouchers Develop and coordinate transportation resources Provide rental assistance Provide UI wage matches Work with One-Stop system

SOURCE: Compiled by Fitzgerald.

Intermediaries are necessary to sustain workforce development networks. Intermediaries act as "social entrepreneurs" who create a vision of a system to guide groups whose participation is needed (Clark & Dawson, 1995). The intermediary role can be played by different institutions or organizations—from community-based organizations, community colleges, or universities to trade associations. Intermediaries envision, plan, implement, and adapt workforce development networks.

Intermediaries serve two important functions. First, they make connections. It is not always obvious to potential partners that they can assist each other in achieving their organizational goals. Second, intermediaries resolve conflicts that emerge among partners. Although some degree of competition can be useful, in many cases it prevents organizations from seeing the potential for partnership. SJI itself is an overarching intermediary that facilitates and coordinates a network of stakeholders in economic and workforce development ranging from state government to labor unions. Each targeted sector has a team that also serves as an intermediary. Each targeted sector team is composed of a sector manager, a broker, CBOs, and training providers.

Sector managers coordinate all of the workforce development functions into a coherent program and coordinate biweekly meetings between brokers, CBOs, and training providers. The meetings provide an opportunity to discuss how SJI is meeting the needs of both participants and employers and identify needed program changes. As these meetings evolve, sector managers decide whether less frequent meetings or other types of communication would allow the team to handle business more effectively.

Brokers are the key link between employers and clients. As the liaison among employers, training providers, and CBOs, brokers are chosen based on their strong ties to the industry. The brokers work with employers experiencing labor shortages, high turnover, or both to identify workforce and training needs. They also work with the CBOs that refer candidates for job openings to make sure they understand employer needs. In addition, the brokers involve employers in curriculum development and encourage them to provide internships. The brokers vary by sector and range from well-connected individuals to industry associations such as the Washington Aerospace Alliance in manufacturing.

The use of brokers is a new approach in workforce development. Education and training providers offer varying levels of placement assistance to students, but placement specialists rarely have the strong ties to employers that brokers have. The role of the brokers has evolved over time. Some initial misunderstanding emerged between SJI staff and the brokers on their role. SJI staff assumed the brokers would work with employers to identify opportunities for entry-level positions for participants. In many cases, employers were reluctant to hire welfare recipients or applicants with other perceived disadvantages. SJI staff expected the brokers to educate and sensitize employers on issues of diversity and hiring workers who may not fit their image of the ideal employee. The brokers, however, thought of employers as customers whose needs they were hired to satisfy and expected the CBOs to send only clients who fit the expectations of employers to interviews. If an employer requested specific racial or ethnic preferences, these were simply passed on to the CBOs by the brokers. The brokers were frustrated that many of the clients the CBOs were sending were not what the employers requested. The CBOs were frustrated that their clients were not getting hired.

To resolve this problem, SJI staff asked the brokers to serve more as a liaison between the CBOs and employers. The brokers had to change their own and employer attitudes about the target population and convince them to try new employees who may not fit their ideal expectations. The issue was more than employer attitudes; some of the clients CBOs recommended for particular jobs were inappropriate. Thus, the brokers began working with the CBOs on improving their assessment of clients before recommending them for placement. The brokers now keep CBOs informed on how referrals are working out with employers and alert CBO case managers about problems so they can intervene to help clients stay employed.

The brokers also work with community colleges to determine whether existing courses fit the needs of participants and employers. In some cases, SJI participants enroll in regularly offered courses; however, mostly special classes are offered. As the brokers work with employers to learn their level of satisfaction, they return to the community colleges with requests for curriculum changes. This role has become increasingly important in ensuring that curriculums meet employer needs. In one case, a community college's students in an administrative office skills training program earned certificates but were frequently unable to pass employer screening tests. The brokers worked with the community college to develop a competency-based curriculum, to bring on employers as instructors, and to create internships for students.

Through implementation, SJI staff are learning the ideal division of labor among the brokers, CBOs, and community colleges and are continually adapting their responsibilities. The key role the sector teams play is ensuring that each partner is doing its job. As intermediaries, the brokers have had to change attitudes and practices of employers, CBOs, and education institutions.

Effective initiatives integrate poor people into recognized education and training programs. This principle defines the distinction between a collection of programs and a system. A sectoral workforce initiative allows for the creation of industry- or occupation-targeted education and training programs that have several points of access and several types of programs to serve different client groups. This approach contrasts sharply with most federal job-training programs.

A key reason for the failure of so many past federal employment training programs, including Manpower Development Training Assistance, Comprehensive Employment Training Assistance, and the Job Training Partnership Act, is that by focusing overwhelmingly on the poor, they stigmatized clients in the eyes of employers (Osterman, 1988). Too often, the disadvantaged are placed in programs for which there is little demand and in occupations offering low wages and little opportunity for advancement. With few exceptions, there is little evidence to demonstrate that these programs have done much more than provide subsidies for filling low-wage openings. Channeling of the poor into inferior programs is also evident in community college adult basic education (ABE), general equivalency diploma (GED), and English as a Second Language (ESL) programs. These programs expanded rapidly in the 1980s and continue to enroll the majority of students in many urban community colleges. Yet few community colleges have been successful in moving students from basic skills classes into credit or even certificate programs.[8] One solution is to create bridge programs so those who are not ready for college level work, whether certificate or degree programs, can acquire the academic skills needed to complete them (Jenkins, 1999). These can be offered by CBOs or the community college. Identifying this gap in the education system as an economic development issue opens possibilities for economic development practitioners to act as intermediaries who encourage local training and education providers to be more responsive to the needs of key local employers.

The SJI, by working with community colleges to create bridge programs, is taking on this role. After discovering that only 50% of clients were completing a computerized numerical control (CNC) class at Shoreline Community College, SJI staff analyzed the situation and found the clients were lacking in both hard and soft skills needed to complete the course. The solution reached by the Aerospace Sector partners was to improve the academic preparation of clients by arranging with the Seattle Vocational Institute to offer a Vocational Adult Basic Education (VABE) program that combines instruction in basic math with blueprint reading. The VABE program provides a bridge to community colleges for those who need to build academic skills. This approach of combining literacy and specific job skills is considered cutting-edge practice in community colleges throughout the country (see Fitzgerald & Jenkins, 1997). One of the CBO partners in the Aerospace Sector now provides a week-long pretraining job preparation and readiness program.

SJI employs other approaches for making community college curricula more responsive to the needs of underprepared students and employers. In all training programs, SJI ensures that basic and soft skills training are integrated into the curriculum, that the training environment replicates the work environment, and that employers are involved in developing a competency-based curriculum. When a community college offering classes for the Business-Office Occupations strategy would not customize a class for SJI clients, a partnership was formed with the Seattle Vocational Institute, which was willing to tailor its programs to meet employer needs. This arrangement requires having community college staff willing to invest time in making curriculum changes and SJI having enough students to place in the classes. Because community colleges recognize that SJI has become a key workforce development intermediary, they are motivated to be more responsive to requests to update and adapt their programs. In this intermediary role, SJI is increasing access of low-income residents to higher levels of education and ensuring that local educational institutions are responsive to the needs of employers.[9]

Effective workforce development systems focus on improving employment retention and identifying opportunities for advancement. These are the two biggest challenges in workforce development. Most residents of low-income communities qualify only for entry-level, low-wage jobs characterized by high turnover, few benefits, and little opportunity for advancement (Bernstein & Mishel, 1995; Blank, 1997). The issues of job retention and advancement are highly interrelated. Low wages are associated with low levels of labor market attachment. A longitudinal study of welfare recipients in the state of Washington from 1988 to 1992 found that the wage levels received by former welfare recipients relate to the length of time they remain off assistance (Lidman, 1995). Of those welfare recipients who were earning hourly wages of $9.50 or more, 67% remained off assistance 3 years later. Of those who earned less than $6.50 per hour, only 32% remained off assistance for 3 years. Analysis of employment patterns from the National Longitudinal Study of Youth also reveals that once women find good jobs, they stay in them regardless of characteristics such as former welfare use, education, race or number of children (Pavetti & Acs, 1996).

SJI is experimenting with several strategies for improving job retention. One effort takes place before training even begins. After finding many participants quit jobs because they didn't know what to expect and didn't like the work, SJI developed pretraining sessions that explain what individuals can expect in an occupation. Participants who have a good sense of what a job entails before training are more likely to stay on the job. Once employed, participants are eligible for postplacement services for up to 2 years.

Another retention strategy is peer mentoring. The Men of Action Group was created as a peer network to retain clients in school and jobs. The group was created because many African American clients enrolled in the CNC program at Shoreline Community College were uncomfortable in its suburban environment and also in the almost exclusively white work places in which they were placed. The CBO partner in the Aerospace Sector Strategy coordinated the Men of Action Group to provide a forum for the men to identify ways to overcome racial barriers and to discuss work and life issues that affect their ability to maintain good attendance and working relationships on the job. A 12-week life skills curriculum is offered through the weekly meetings. Further, case managers provide diversity training for employers and mediate disagreements between supervisors and workers.

The most pressing retention issue is the low supply of jobs paying the $8 hourly wage required by the initiative. There is little motivation to stay in a low-paying job when transportation, child care, or both are problematic and few opportunities for advancement exist. Workforce development systems, in addition to addressing individual barriers to employment, have to work with employers to improve the quality of jobs. One approach is to develop career ladder programs that enable low-wage workers to advance through a progression of higher-skilled and better-paying jobs (see Fitzgerald & Carlson, 2000). To this end, the SJI is examining the extent to which career ladders can be integrated into training programs. A promising new program is Tech Talent, which provides training and internships for the position of junior web designers with possibilities for moving up to master web designer.

The Seattle OED has a staff liaison who works with high-tech employers, mostly on space and infrastructure needs. In the course of these interactions, several employers mentioned a shortage of workers in web design, network administration, and other technical positions. SJI staff had already been thinking of developing a training program in web design and had traveled to San Francisco to talk with staff from OpNet and the Bay Area Video Coalition, two organizations operating highly successful web design training programs for low-income groups. These programs offer 6 weeks of full-time classes followed by a 6-month paid internship. The OED liaison discussed the labor market shortage with SJI staff and joined the high-tech sector team to explore creating a training program. A broker with experience in the field was hired to talk with Human Resource staff in high-tech companies to identify the specific qualifications needed for web design positions. The employers were hesitant to consider hiring people without college degrees, but because of difficulty in finding qualified applicants, they were willing to consider hiring graduates if SJI developed a program that provided all the skills and tasks they had identified for web designers. SJI approached South Seattle Community College about developing a program along the lines of the San Francisco models. The college staff compromised by offering a version of its 1-year program as a 22-week, full-time program (8 a.m.-5 p.m.) followed by a 3-month internship. SJI is trying to convince the three partner companies to offer paid internships.

Twenty-three people were accepted into the first class, which started in June 2001. Neither a high school diploma nor a GED is required, but applicants must be literate in math, writing, and reading at the 10th-grade level and know basic keyboarding. Graduates will start in jobs offering between $12 and $15 per hour and can advance into higher-level positions. Tech Talent is the first SJI program that places participants in jobs that truly offer a living wage and have career advancement potential.

Job-Centered Economic Development in a Suburban Context

As discussed in the book's Introduction, poverty and unemployment are increasingly becoming suburban problems. Nationally, there is a pressing need for services to the unemployed in many suburban areas. In the Seattle area, this need has emerged in southwest King County. Due to rapidly rising housing values, poorer people are being forced out of Seattle into southwestern King County. The King County Jobs Initiative (KCJI) was developed in response to their employment needs.

The KCJI was formed in early 1998 to assist low-income residents in finding jobs paying at least $8.00 per hour. An interdepartmental team formed by the King County executive's office developed a plan for the initiative. The Republican-controlled Metropolitan King County Council wanted assurance that it would be an economic development program, as opposed to a social work program, before approving $1 million in startup funds. Thus, the language framing the KCJI is focused more on meeting the needs of employers for skilled workers than the need for living-wage jobs as in the SJI. The program is operated by the King County Office of Regional Policy and Planning.

The KCJI follows the same basic principles as the SJI. Several sectors have been targeted and relations established with the community college serving the three suburbs served by the KCJI. Four CBOs provide recruitment, placement, and support and retention services. Rather than using the terminology of social work—case managers—the clients are served by account executives. In addition to traditional case management activities, the account executives are also responsible for job development. An employer liaison facilitates employer involvement and job placement.

The smaller scale of the KCJI means that it does not have the same leverage to arrange customized classes with community and technical colleges as SJI. Nevertheless, the KCJI has influenced community college practice. Realizing that students served by account executives are more likely to complete their programs, South Seattle Community College negotiated with state and county decision makers to become a One-Stop Center affiliate.

KCJI staff are working with the college in developing an innovative program in environmental cleanup, a set of occupations offering well-paying jobs with career ladders and steady demand in the Seattle area. KCJI and employers in this sector are working with South Seattle

Community College to update its curriculum in this field. Funds for program planning and implementation were provided by a $147,000 grant from the Environmental Protection Agency, which jointly designated the City of Seattle and King County as a Brownfields Showcase Community. The educational innovation is that courses in the Brownfields Job Training Project are organized in 8- to 24-hour modules, with each module counting toward an environmental technician associate degree. Typically, short-term certificate courses do not provide credits toward a degree. Each module qualifies the graduate for certification in a specific type of cleanup. Instruction covers both technical and safety concerns. Graduates of the first environmental cleanup module are qualified to work in disposal of hospital waste, cleaning of contaminated sites such as shipyards, and recycling of hazardous materials such as paint. With each module and certification completed, participants can earn higher hourly wages. Workers start at $11.50 per hour, and associate degree graduates in the field earn between $30,000 and $40,000 and in some cases considerably more. KCJI is targeting women to fill these relatively high-paying jobs. Other modules include Innovative Technologies and Advanced Environmental Assessment.

Can Job-Centered Strategies Create Systemic Change?

SJI and KCJI are working to increase standards of living and reduce inequality among Seattle and suburban King County residents. Improved living standards are achieved by placing people in jobs to which they normally would not have had access. These initiatives are laying the groundwork for systemic change in workforce development in four areas:

Developing, Implementing, and Documenting Successful Program Practices That Can Be Replicated. The SJI and KCJI have influenced organizations that compose the workforce development system to be more responsive to the needs of low-income clients. Although the SJI does not have enough clients in many sectors for community colleges to create customized programs, the brokers have convinced three community colleges to revise their curricula to be more competency based and responsive to business needs (see Fitzgerald, 1998 on this issue in other sites). CBOs have changed the services they provide to their clients and are more responsible for making sure the clients are job ready when

TABLE 7.3   SJI Placements

Project or Sector	Placements Through 5/01
Automotive	61
Diversified manufacturing (CNC)	69
Diversified manufacturing (electronics)	84
Diversified manufacturing (general)	33
Office occupations	207
Individualized placements	1,523
Employer response (no services)	80
Conservation Corps	14
Health care	68
Prep employment	69
AOP family wage placement	68
Apprenticeship placement	68
TOTAL	2,344

SOURCE: Adapted from Seattle Jobs Initiative.

placed. Brokers have had to learn to work more closely with CBOs. Businesses are hiring workers they previously would not have considered. The SJI and KCJI partners are creating a coordinated continuum of services—from basic placement to postemployment counseling that provides a structured environment for those making the transition from unemployment to work. The link between workforce development and human service delivery has been more effective than job training or placement alone.

Another aspect of systemic change is documenting outcomes. SJI's long-term job retention and advancement goals are difficult to measure. SJI and the other five Jobs Initiative sites are working to create an operational definition of job retention and a cost-effective system for tracking clients. The bottom line, however, is placements. Although SJI has placed 2,344 people since it started, 65% are through direct placements rather than through the sectoral initiatives (Table 7.3).

Over time, the placement and retention rates of the sectoral projects have improved. Projects have been dropped (e.g., health care) when placement rates, retention rates, or wages were low. As the sectoral projects become more established, their retention rates are moving beyond those of direct placements. Currently, 54% of direct placements are working 1 year after placement. The corresponding figure for office occupations, electronics assembly, and apprenticeships are 66%, 67%, and 60%, respectively. This suggests that sectoral strategies take several years to establish.

Advocating for Improving and Integrating State Workforce Development, Economic Development, and Welfare and Human Service Policies and Programs. The SJI and KCJI, working with Seattle's mayor and the King County executive, have influenced Washington's workforce development policy to support the goals of job-centered economic development. SJI and KCJI staff have been part of a work group appointed by Governor Locke to make recommendations for reforming the state's workforce development system under the Workforce Investment Act. In this capacity, they are advocating for strong local control of workforce boards. Further, the SJI was able to influence how One-Stop Employment Centers are organized in Seattle. In addition to Seattle's three One-Stops designated by the state, SJI fought for and won an additional decentralized One-Stop operated by CBOs and for CBOs to be eligible One-Stops service providers and eligible points of entry to the One-Stops system, which will make the system more accessible. The connection to the One-Stops is essential for creating systemic reform because they are the core delivery mechanism for job placement and federal employment training programs.

Job retention and advancement are two additional areas in which the SJI and KCJI are attempting to influence state policy. Like many states, Washington has experienced considerable savings through declining welfare rolls. Seattle's Mayor Paul Schell, King County Executive Ron Sims, and others have advocated for the governor to invest the savings from declining welfare rolls into education and training for WorkFirst clients. Several initiatives have passed as a result. Governor Locke issued an executive order to the State Workforce Board to create wage progression and advancement strategies. The state's community college system has received additional funds to develop wage progression and career ladder programs. The state transferred $17 million from the Department of Social and Health Services (DSHS) to the State Board for Community and Technical Colleges in 1998 for developing programs to promote job advancement and wage progression.

In 1999, an additional $20 million was allocated for three programs. The first is a 12-week pre-employment training program that links to the needs of one employer or group of employers. A work-based learning program provides tuition assistance to serve as a bridge between free tuition and eligibility for federal Pell grants. Any parent under 175% of the poverty line and working 20 hours weekly is given free tuition to any community college technical program, usually for one or two quarters, until Pell eligibility kicks in. A third program designates funds to community colleges for redesigning programs, adding more certificate programs, and offering more evening and weekend courses to make it easier for students to combine school and work.

In 1999, a work-study program was added for TANF recipients that provides part-time employment, usually with community-based organizations, for students enrolled in college courses. Work-study jobs must be related to the student's course of study and can last no longer than two academic quarters.

This is not a claim that the programs would not have been created but for the influence of SJI and KCJI, but certainly the initiatives have been key players in influencing state and local workforce development policy. The environment of cooperation between state and local government and the SJI have made Washington a leader in workforce development policy (see Massing, 2000).

Encouraging Greater Integration of Regional Employment and Training, Economic Development, and Human Services Policies. As more suburban employers experience difficulty in finding qualified workers, the potential for partnerships between suburban and urban workforce development increases. SJI staff have been involved in a metropolitan-level organization committed to coordinating workforce development policy and influencing state policy to better serve the needs of low-income populations in Seattle and its surrounding suburbs. In 1997, SJI partnered with several workforce development organizations in King County to form the Coordinated Funders Group. This partnership of funders defined its mission as establishing a regional and comprehensive training, employment and retention system that provides efficient and effective services to employers and low-income job seekers and workers in King County. An interest of the group was to shift the focus of welfare reform from "work first" to training. The group was a voice in advocating for the state's investing savings from reductions in the welfare rolls in job training.

The child care subcommittee of the group successfully advocated for increasing income cutoffs for eligibility for job-training funds and subsidized child care from 175% to 200% of the poverty rate. The transportation subcommittee created a transportation broker position to work with employers and the Regional Transportation Authority to start van pools. An ad hoc committee developed wage progression strategies that influenced the state wage progression funding reallocation mentioned earlier. Finally, King County members of the group formed the KCJI to serve the training needs of suburban King County residents. These efforts illustrate the group's focus on moving beyond a placement focus to one that attempts to move people out of the ranks of the working poor.

Influencing the Behavior of the Private Sector Through the Targeted Sector Strategy. The SJI and KCJI have had considerable success in creating partnerships with employers in key sectors and being responsive to their needs. The weakness of sectoral strategies, however, is revealed when major employers experience a sudden shift in demand. Boeing's intermittent layoffs over the past several years have reduced employment opportunities. When SJI started the program, there was considerable demand for CNC machine operators among small- and medium-sized machine shops that had been losing skilled workers to Boeing. The 12-week CNC training program helped prepare participants for these jobs. As Boeing started laying off workers, demand dropped in the smaller firms, and the 12-week program was no longer sufficient to qualify participants for available jobs.

SJI staff are aware that being "employer driven" does not take workforce development far enough in achieving systems reform. At some point, SJI has to convince more employers to pay better wages and offer more advancement opportunities. SJI staff are finding that there is an increasing amount of contingent work in the local economy, making it more difficult to place clients in permanent jobs paying $8 per hour. As welfare reform continues, the SJI is serving a population with more barriers to employment. Most of the entry-level jobs they are placed in have few opportunities for advancement.

The SJI is pursuing a gradual strategy of establishing legitimacy with manufacturing employers by supplying dependable workers and then approaching them about career ladders or paying better wages. The initiative has not yet had much influence in this regard. Efforts to create job ladders in the targeted sectors have not been met with employer enthusiasm. An attempt to create an upgraded training program in electronics assembly for working graduates of the 12-week certificate program was not successful. The sector manager and community college staff invested a considerable amount of time in developing the program, but state funding was not approved.

Lessons for Local Economic Development Practitioners

Seattle and the other five Jobs Initiative cities are in a unique position in having a well-funded 8-year period to plan, develop, and implement an approach to workforce development that seeks to make existing systems more effective and equitable. As with most demonstration programs, the hope is that other communities can learn lessons and adapt their systems accordingly. Seattle has been able to leverage funds from the Annie E. Casey Foundation into additional support from government agencies, corporations, and other foundations. This broad funding base allowed SJI to expand from an operating budget of $3.7 million in 1997 to $6 million in 1998. The Seattle City Council created the Families and Jobs Opportunities Fund, adding an additional $5.3 million for implementation of the first two phases. Yet the fact that the KCJI has been successful using the same basic approach suggests that the approach can be replicated without a similar infusion of funds.

An essential skill for economic development practitioners is the ability to recognize when things are not working and to develop new programmatic responses. The ability to adapt has been essential to the success of the SJI and KCJI. Examples of changes made include dropping sectors when labor market demand declined, redefining the roles of partners and intermediaries, and expanding the initiatives by developing new partnerships with agencies in King County and the state.

The success of an economic development initiative often depends on organizations over which the practitioner has no control. As intermediaries, SJI and KCJI are trying to change the practices of organizations involved in workforce development. For the most part, however, the funding provided by SJI and KCJI to any one organization is not enough to influence it to change practices dramatically. In the case of community colleges, for example, the number of students provided by SJI and KCJI usually has not been enough to justify customized classes or revamping the curriculum.

The relatively low number of placements in the targeted sectors compared with direct placements could just reflect that it takes several years to identify sector needs, put partnerships in place, and gain the trust and cooperation of employers. Indeed, placements through the sector initiatives are increasing over time. The low numbers could also suggest that an exclusive focus on sectors, rather than occupations, may be misguided. Anne Keeney, SJI's sector manager, points out that "the labor market can shift suddenly and you need the flexibility to place people in a variety of environments with the same skill set" (personal interview, June 2001). This is particularly true in high-tech occupations such as web designer, where graduates could be placed in several sectors.

The SJI and KCJI have been catalysts in creating an effective link between local and state workforce development agencies. Because WIA puts the power of implementation into the hands of state government, achieving systemic reform geared toward moving people out of poverty has to be a goal of both local and state officials. In Seattle, this goal has been embraced by two mayors and key staff implementing the initiative and in King County by the county executive. The good relationships between local actors and state government agency staff have allowed them to negotiate how federal legislation (WIA and welfare reform) is implemented to achieve more equitable outcomes.

Both initiatives are aided by strong connections to local government economic development departments. SJI is part of the Seattle Office of Economic Development, and the staffs of both groups are in frequent communication. Economic development staff inform SJI when they identify workforce needs and also work in partnership with SJI on developing training initiatives. The Tech Talent program illustrates the kind of synergy that exists in meeting the economic and workforce development needs of businesses in targeted sectors. As an organization of the King County Office of Regional Policy and Planning, the KCJI also links its programs to the economic development priorities of the county. In the late 1990s, several suburban King County cities hired economic development managers, including Bellevue and Renton, which are served by the KCJI (Erb, 1999). Time will tell how these managers link their development efforts to the KCJI.

The definition of economic development offered in Chapter 1 includes the goals of increasing incomes and reducing income inequality. As defined here, workforce development seeks to increase the quality of jobs in the local economy and to increase the access of poor people to these jobs. There are two barriers to achieving these goals. First, many of the jobs being created, even in a rapidly expanding economy such as Seattle's, are low-wage, low-skill jobs. Employers have not expressed much interest in efforts to encourage them to create more advancement opportunities. Second, the federal legislation influencing the ability of practitioners to achieve these goals defines success simply in terms of placing people in jobs. By funding service providers for working with clients for up to 2 years, SJI hopes to emphasize the importance of retention and advancement, but as discussed previously, SJI has limited leverage. Influencing the demand side is the ultimate test of job-centered economic development. The SJI has several years left to make gains in this area.

Notes

1. States are permitted to exempt 20% of their caseload from the time limit.

2. The other cities are Denver, Milwaukee, New Orleans, Philadelphia, and St. Louis. The criteria for selecting the sites were a strong civic base, an organization willing to take responsibility for planning and implementation (a development intermediary), an existing track record of jobs projects, and local philanthropic support to match the foundation investment.

3. Approximately one half of SJI clients are within this age group.

4. First-source hiring agreements require employers receiving public subsidies or incentives to give priority consideration to hiring residents of the jurisdiction offering the subsidy. In some cases, employers are encouraged to consider TANF recipients or other hard-to-place populations.

5. The discussion of changes in sectoral initiatives draws from the *SJI Revised Strategic Plan* (City of Seattle, Office of Economic Development, 1999).

6. Soft skills refer to knowledge of appropriate work behavior such as showing up on time, dressing appropriately, communicating clearly, etc.

7. In a study of New Jersey's REACH program, Hershey and Pavetti (1997) found that 43% of clients cited child care, along with health problems, pregnancy, and family problems, as key reasons for losing jobs. They also argue that women with formal child care stay employed more consistently than those with informal arrangements.

8. See Fitzgerald and Jenkins (1997) for examples of community colleges that are effective in mainstreaming the poor into credit programs.

9. SJI, for example, has been able to pressure community colleges into moving toward more competency-based curricula in CNC, business, and office training.

8

Strategies and Progress
for Local Economic Development

Paths to Improving Local Economic
Development Planning and Practice

Our first goal in writing this book is to examine how economic and social equity and environmental sustainability can be built into common economic development strategies. We identified three principles that underpin an economic development practice focused on economic equity and sustainability: increasing standards of living, reducing inequality, and promoting and encouraging sustainable resource use and production. To incorporate these principles into practice, we proposed an alternative definition of economic development to that of the conventional business interest-driven and wealth-creation approach currently dominating the field. In our view, this conventional approach errs in seeking economic growth over economic development. We make a distinction between the two: Economic *growth* is more development, more businesses, more jobs, and more taxes, whereas economic *development* is raising standards of living and improving the quality of life through a process that specifically lessens inequalities in metropolitan development and improves the metropolitan population's standard of living. This distinction between growth and development is not oriented solely to the present because economic *development* is *sustainable*. It is growth and change that neither contribute to rising inequalities nor diminish opportunities for future generations. Thus, as we stated in Chapter 1: *Local economic development builds a stable economic base that preserves and raises a community's standard of living by developing its human and physical infrastructure in a sustainable and equitable manner.*

Our second goal is to discuss how common urban economic development strategies can be applied in older and inner-ring suburbs in addition to cities. This focus is essential because continued suburbanization and exurbanization of population and economic activity reinforces unsustainable resource and land use practices at the same time that it widens inequalities and leaves even farther behind those living in the inner cities and inner-ring suburbs. Examining economic development problems through the lens of the inner-ring suburb and the inner city helps to make clearer what our nation's greatest local economic development challenges are and provides the needed focus for resolving them. Although our book is aimed at helping the field of local economic development become more effective in solving the problems of inner cities and inner-ring suburbs, we believe that our discussion of the strategies can also help practitioners in thriving new towns and suburban communities prevent economic decline as their communities mature. Chapter 4 on industrial retention identifies a few instances where inner-ring suburbs are focusing on industrial retention but suggests that more should be paying attention to industrial migration. Chapter 6 on the reuse of office and industrial properties seeks to give economic development students and practitioners an understanding of some of the possibilities for creating new economic activity in the built environment of left-behind inner-ring suburbs and inner cities. Chapter 7 on workforce development details how an urban strategy is now being used very effectively in inner-ring suburbs.

The third goal is to inform the teaching and practice of economic development by focusing on implementation and the politics of local economic development. We have provided an in-depth focus of implementation in our chapters on industrial retention, sectoral development, and workforce development strategies to enhance understanding of how the planning process works. The cases of implementation in these chapters reveal the political nature of the planning process and the types of trade-offs on issues of equity and sustainability that often must be made. We also focus on implementation because of the insights it reveals to planners seeking to adopt "best practice" programs from other localities. In reality, best practice programs are the result of frequent adaptation over months or years of implementation. With an enhanced understanding of the implementation process, economic development practitioners can better adapt strategies such as those featured in the industrial retention, commercial revitalization, brown-field redevelopment, and workforce development chapters to fit the needs and political realities of their locality.

Evaluating the Strategies

We turn now to a more in-depth discussion of how the economic development issues raised and goals we set in Chapter 1 are satisfied in the six specific strategies presented in this book. As we stated in the Introduction, we did not expect that each strategy would be able to satisfy all of the criteria we set for economic development planning and practice. Rather, we profile only those local economic development strategies whose implementation at least partially moves the community closer toward achieving what we define as economic *development*.

We evaluate the strategies by translating the goals, principles, and definition into four discussion points: (1) their ability to incorporate equity goals, (2) their ability to incorporate sustainability goals, (3) their transferability or relevance to inner-ring suburban settings, and (4) issues encountered in implementation, including trade-offs among equity, sustainability, reducing inequality, and raising standards of living; political constraints to taking or planning action; and how legislation limits or creates opportunities for incorporating equity and sustainability.

Incorporating Equity

Equity, defined by the goals of creating economic development opportunities for poor people or minority populations, is evident to some degree in all the strategies presented. In some cases, such as sectoral strategies, brownfield redevelopment, and workforce development, equity is a principle defining the strategies. In industrial retention, commercial revitalization, and office and property reuse, it takes concerted action on the part of practitioners committed to this goal to see that it is incorporated into the strategy.

To the extent that sectoral strategies provide access to living-wage jobs for people who normally would not have access to them and convince employers to create more advancement opportunities and pay better wages, they promote equity. The two sectoral strategies presented have different emphases on this goal. Jane Addams Resource Corporation provides training that allows incumbent workers to move into better-paying positions in manufacturing. The program has been extended to low-income residents, who are now taking entry-level positions from which they can advance into better-paying jobs. Although there are some employment opportunities for people with associate degrees, the biotechnology sectoral strategy mainly benefits people with advanced degrees. The City of New Haven is, however, including light manufacturing in its high-tech park to create some employment opportunities for residents in neighboring low-income communities.

Does this mean that we should support one strategy over the other on equity grounds? If the high-tech sectoral strategy is the only strategy being pursued, it may represent a trade-off, although not necessarily an inappropriate one. For example, biotechnology, at present, is largely seen as a highly advanced, leading-edge industry employing primarily research scientists and engineers. Yet in New Haven, CuraGen discovered that lab workers with associate degrees were quite capable. Further, California's Employment Services Department has identified a range of lower-skill jobs that can be found in the industry sector including animal handlers and lab technicians (California Trade and Commerce Agency, 2000). In Kennesaw, Georgia, the biotech firm Cryolife hires employees with vocational degrees or only high school degrees to work as lab technicians. They are trained in-house, starting with an intensive 3-month program and continuing for 2 years (M. Schoenberg, personal interview, October 22, 2001). The firm harvests heart valves, knee cartilage, and veins from donors for transplant operations. The training program provides the workers, many of whom are female and minority, with the required skills to do the harvesting. Cryolife offers above-average wages and a very complete benefits package, including stock options and tuition reimbursement to any full-time employee. The lowest beginning wage is $10.50 for those with high school degrees, and employees hired at this level receive significant increases within their first year of employment.

These opportunities suggest that blindly criticizing high-tech sectoral strategies on equity grounds may well be shortsighted. Rather, the focus should be on creating career advancement opportunities for entry-level workers. Employers may take some convincing that this is a viable strategy, as we saw in the case of the Seattle Jobs Initiative. However, these examples suggest that it is possible with sustained effort.

Equity is a key focus of Chapter 3 on brownfield redevelopment. Currently, the strategy, as it is predominantly practiced and promoted by public and private sector entities alike, is market based. As such, the chapter criticizes most current brownfield redevelopment practice for subordinating considerations of equity to those of pursuing projects that will pay for themselves. This means the focus is on the most marketable brownfield sites that are typically not found in the lowest-income areas of either inner-ring suburbs or inner cities. However, the chapter also highlights the efforts of communities that specifically are incorporating issues of equity in their brownfield redevelopment strategies. These cities—Kalamazoo, Michigan, and Louisville, Kentucky, for example— pursue brownfield redevelopment within the overall context of community revitalization. The chapter also profiles public sector efforts to secure the funds necessary to pay for non- or less marketable brownfield sites in need of redevelopment, a necessary component of an equity-focused brownfield redevelopment strategy.

Industrial retention is an equity strategy to the extent that it promotes retaining high-wage jobs for those who have lower-skill levels. The Chicago case study reveals that although the industrial retention "cause" was started by a progressive city administration for which equity was a primary consideration, retaining industry has become so much a part of the culture of economic development in the city that it has been a priority of subsequent mayors. This approach is not without its costs, as illustrated in the ongoing political debates over land use near the Loop and in the large subsidy just provided to Ford to develop a supplier park. The question of whether the development would have been possible with less public subsidy always remains. Further, as more inner-ring suburbs take on industrial retention, they will lack the resources to offer subsidies that large economically diverse cities like Chicago have.

The commercial revitalization strategy in Chapter 5 satisfies the principle of equity by seeking to provide inner-city and inner-ring residents with equity in access to good consumer choices, fair prices for goods and services, and commercial sector employment opportunities. Inner cities and inner-ring suburbs have untapped retail potential that is a polite way of saying that the retail sector has been underserving them or avoiding them. Thus, residents spend their retail dollars outside their areas, when they have access to transportation, enriching the retail economies of wealthier areas. Pursuing commercial revitalization will help to reverse the retail spending leakages and promote stronger local economies.

The office and industrial property reuse strategy profiled in Chapter 6 makes the least obvious contribution of the book's six strategies to the equity principle. We know that the demand for reuse of office and industrial properties comes primarily from the private sector, and that successful private developers are profit maximizers, operating on the premise that much of a potential reuse project's final value will be determined by its location. Thus, properties located in a gentrified area, or area with high potential to gentrify in the near future, will be those focused on by private developers. Unless efforts are made to promote the reuse of industrial and office properties in the most marginal or least-likely-to-gentrify areas, the very act of redeveloping and reusing such properties in their strongest areas will serve to create greater inequities within older cities and suburbs. To promote equity, local economic development practitioners can seek opportunities to invest public development dollars in the reuse of buildings in the areas the private sector will avoid, essentially following a local version of the federal policy to site federal buildings in the central cities.

The Seattle and King County Jobs Initiatives represent a broader movement to incorporate equity principles into workforce development. Although the nation's history of job-training programs has mostly been one of targeting low-income populations or displaced workers, its overall effect in moving people out of poverty has been minimal. In contrast, the job-centered approach to economic development sets living-wage work and career progression as goals. In both cases, low-income populations have increased access to jobs paying a living wage, defined at the relatively low rate of $8.00 per hour. However, the initiatives are finding it difficult to find enough jobs at even this rate and are only just beginning to convince employers to create career advancement possibilities. These examples and others illustrate that economic development practitioners have very limited ability to affect local labor markets.

Incorporating Sustainability

Environmental sustainability is at the core of brownfield redevelopment, industrial retention, and office and industrial reuse strategies.

Sustainability can also be incorporated into commercial revitalization and sectoral strategies, as illustrated in our cases. Workforce development can be integrated into any of the economic development strategies presented in this book and thus can be linked to sustainability goals.

In both the metalworking (Jane Addams Resource Corporation) and biotechnology sectoral strategies, abandoned buildings in inner cities were purchased and renovated into manufacturing or laboratory space. In so doing, they either stemmed further suburbanization and exurbanization of manufacturing activity or made the city a viable location for an industry that has predominantly located in suburbs. In each case, available buildings and infrastructures were reused rather than developing greenfield property.

At the core of the brownfield redevelopment strategy is promoting sustainability through leveling the playing field between greenfields and brownfields for all types of development. When this occurs, there is less impetus to consume greenfields, permanently altering their undeveloped status. Instead, the brownfield redevelopment strategy removes the barriers to reusing previously developed properties for new and expanding economic activity.

Because industrial retention efforts make it possible for manufacturers to expand in existing facilities rather than moving to greenfield sites, they contribute to sustainable development. Portland and Seattle have been able to accomplish this through their industrial sanctuaries and manufacturing industrial centers. In both cases, concentrating industrial uses is part of a broader regional growth management strategy. Chicago has retained several manufacturing firms because of the planned manufacturing districts, but there is still considerable development of manufacturing on greenfield suburban sites. In other retention efforts, Chicago has linked maintaining the Ford assembly plant with construction of a supplier park in an abandoned brownfield area. Ford chose this option over a greenfield site it was considering in suburban Atlanta. Further, maintaining industrial areas within the city preserves jobs for residents in nearby areas. If these jobs were relocated, employees would have to either move or commute farther to work, contributing to either sprawl or pollution.

The office and industrial reuse strategy complements that of brownfield redevelopment in promoting sustainability, although there was no explicit intent to do so in the case studies profiled. The economic development practitioner, however, can seek to promote office and industrial property reuse for the broader goal of sustainability, as well as its contribution to economic revitalization and strengthening property markets. Efforts to promote all three objectives are enhanced when, as the chapter case studies indicate, the properties needing reuse have appeal due to their design aesthetics, quality construction, or both. Good architectural and urban design promotes the long-term functional and aesthetic attractiveness of an area's properties, thereby enhancing their reuse and contribution to sustainable patterns of development. If economic developers will keep this in mind as they pursue new and reuse projects, seeking to use planning and urban design tools such as zoning and design standards to shape private and public construction, they will contribute to the long-term sustainability of their community.

This is also one of the key points made in the commercial revitalization strategy chapter. Like brownfield redevelopment and office-industrial property reuse strategies, commercial revitalization promotes sustainability by strengthening previously developed areas currently experiencing decline. Particularly for inner-ring suburbs, as the Sandy Springs case study shows, they also seek to correct the overly auto-dependent pattern of shopping and dining through the introduction of pedestrian and mass transit features. Further, as the Lawrence, Kansas, Wal-Mart project in the chapter illustrates, commercial revitalization can, at the outset, incorporate design features that promote sustainability into its current function as well as anticipate the inevitability of the retail outlet's obsolescence and need for alternative uses.

Workforce development can also be linked to economic development strategies that promote sustainability. The King County Jobs Initiative in suburban Seattle is training workers in hazardous waste removal. Many of these jobs are associated with brownfield redevelopment efforts. The job training emphasizes worker safety, and the jobs offer relatively high wages for people with moderate skill levels. In Chicago, the Jane Addams Resource Corporation's incumbent worker and other training programs are a key factor in keeping manufacturers in the city. Worker training is a big part of the incentive package that Chicago and the state of Illinois put together to ensure that a brownfield redevelopment site would become a supplier park for Ford. This development will create 1,000 well-paying union jobs for Chicago residents.

Applicability to Suburban Settings

The six economic development strategies can all be implemented in the older, inner-ring suburbs, in addition to the inner city, and are a primary spatial focus of this book. Inner-ring suburbs and inner cities share the same needs for brownfield redevelopment, industrial retention, reuse of vacant office and industrial properties, revitalization of commercial strips and increased access to retail services and goods, workforce development, as well as sectoral strategies to help them develop their comparative advantage in the metropolitan and larger economy. The challenges for inner-ring suburbs can be greater for several reasons. Although inner-ring suburbs can have populations that make them the size of small and medium-sized cities, they are not, for the most part, autonomous political units. Consequently, they must pursue economic development strategies without, for example, the ability to tax, issue bonds, regulate land, and determine their levels of services provision. Instead, they must negotiate with either county or municipal governments to garner the administrative, political, and financial support they require to implement their economic development strategies. This adds time and complexity to the implementation of any strategy.

Sectoral strategies may be difficult for suburbs under a certain size to pursue. Even in a large city like Seattle, some of the sectoral workforce development programs of the Seattle Jobs Initiative do not have sufficient scale to influence employer hiring practices and community college education and training programs. Even in one of the city's dominant industries, aerospace, focusing only on jobs within that sector has made the initiative vulnerable to sudden economic turndowns. Inner-ring suburbs may find it more beneficial to coordinate their efforts with larger regional initiatives. The King County Jobs Initiative illustrates how suburban areas can develop their own sectoral strategies while coordinating with metropolitan-wide initiatives. Likewise, because efforts to develop biotechnology in New Haven are coordinated with a statewide biotechnology strategy, surrounding suburbs are also able to benefit from development.

Many inner-ring suburbs are experiencing loss of manufacturing to more distant suburbs. However, for most, this has not translated into concentrated industrial retention efforts to keep existing manufacturing or to attract new development. This is partly due to the perception that manufacturing is in decline and partly due to the perceived attractiveness of commercial and service sector development. The suburban retention efforts we identified focused on marketing available properties but offered no comprehensive strategies for maintaining strengths in already established industries or creating linkages to strengthen them. This can partially be explained by lack of capacity. Many suburbs do not have economic development staff. Even the largest suburbs in the Seattle area, for example, have hired economic development managers just within the past few years.

In the Sandy Springs case study in the commercial revitalization chapter, the inner-ring suburb created its own revitalization organization to take on many of the tasks that would normally be provided by a planning and development department if the suburb were an autonomous political unit. It arranged for improved policing and street cleaning and maintenance, secured grant monies to fund planning and design activities, and negotiated for a planning overlay district to fine-tune the county's zoning and land use regulations to support commercial and community revitalization efforts.

Inner-ring suburbs' smaller size may also mean that workforce development strategies have to be coordinated with the larger metropolitan area to secure the needed job training and access to appropriate employment opportunities. Inner-ring suburbs are less likely to have a full complement of industrial, office, and commercial activity to provide access to competitive job opportunities in a range of occupations and skill levels. The modification of the Workforce Investment Act that allows suburban communities of 100,000 population or more to operate their own job training, placement programs, or both, rather than participate in a metrowide program, could inhibit access for job seekers in some inner-city and inner-ring suburb communities. This is because of the barriers created by the "jobs-housing imbalance" found in cities throughout the nation where affordable, low-income housing (found most in the central city) is not located in those areas where growth in employment opportunities is occurring (found most in the newer suburbs). Having metrowide access to employment and training opportunities will necessarily require attention to the transportation needs of inner-city and inner-ring suburban residents. Without such attention, the central-city and older-suburb job seekers for whom employment programs are targeted will have fewer choices and opportunities in the labor market.

Issues of Implementation

We identified three issues that repeatedly arose in creating and implementing the economic development strategies covered in this book. The first is making trade-offs. Almost any economic development strategy has supporters and detractors that require compromises to be worked out. Our focus in analyzing the strategies was on whose interests were being advanced and how compromises were achieved. The second implementation issue we discuss is how economic development practitioners gain political support for particular initiatives and build effective relationships for carrying them out. Finally, federal and state legislation places limits on what localities can do in several of the strategies. We focus on the extent to which practitioners can work within the guidelines of existing legislation to advance equity and sustainability goals and how they attempt to change legislation when it creates incentives that increase inequality or reduce sustainability.

Negotiating Trade-Offs

As we have already stated, we did not expect that each of the six strategies would incorporate all the goals identified in our definition of economic development. In almost every case study and example presented, one criterion is emphasized more than another. This is not to say that in each case practitioners make explicit or conscious trade-offs with respect to equity, sustainability, reducing inequality, and raising standards of living. Although this is sometimes the case, other times it is the strategy itself that defines which goals are possible. In yet other cases, the politics of those charged with implementation determines the priority attached to specific goals. Implementing an economic development strategy without encountering political opposition from some groups occurs on the rarest of occasions. Understanding this at the outset allows economic development practitioners to focus on which trade-offs are acceptable and will allow a project to move forward.

The trade-off identified in the sectoral strategies chapter was choosing a strategy. The issue was whether focusing on strengthening old-line industries, such as manufacturing, promotes equity by creating opportunities for lesser-skilled workers whereas strategies that develop new high-tech industries leave many people out and thus increase inequality. Critics suggested that Chicago's manufacturing focus has stunted growth in high-technology industries and San Diego's focus on high technology leaves out the employment needs of the majority of the population. Rather than identifying a trade-off, these criticisms create a false dichotomy between new and old industries and the type of jobs they can create. As we discussed previously, high-tech strategies can create good jobs for workers without a college degree. Clearly, both types of strategies are needed. The dilemma for smaller cities and suburbs is that they need some level of economic diversity from which to build new industries. Further, they may not have locational advantages needed to build all types of high-tech industries.

In the brownfield redevelopment and office and industrial property reuse chapters, the trade-off to be resolved was whether to develop properties in areas of highest need of development or those that would be the most profitable to developers. As the discussion of these strategies makes clear, implementing economic development strategies through only a market-based, private sector perspective has the real potential of widening inequalities between urban neighborhoods and maintaining environmental injustices or inequities. Such strategies trade off projects in which there is the greatest need for redevelopment for projects in which there is the greatest opportunity to engage the private sector in redevelopment. Thus, the practitioner who seeks to foster economic *development* can help to articulate the potential and troubling outcome of possible widening inequality, adjust implementation of the strategy to avoid it, and engage the larger community in supporting the revised strategy.

The industrial retention chapter identifies trade-offs that were made in resolving the conflicting interests of manufacturers and residential developers competing for the same space. The LIRI groups were able to negotiate among the groups in developing the PMDs. The PMD solution, however, came at the expense of new housing development that would have brought more middle-class residents to the city. As Chicago moves toward expanding manufacturing in brownfield sites, there seems to be room for more manufacturing and more residential development in the downtown and close-in neighborhoods. Cities smaller than Chicago in land area may have problems locating all land uses without conflict. However, that does not have to be the case, as revealed in the examples of Portland and Seattle.

Gaining Political Support and Building Relationships

While the preceding section focused on the trade-offs that are often made over strategies and priorities, this section focuses on how relationships are built to create consensus on a common economic development agenda. Trade-offs are often the result of conflict. In contrast, relationship building is not necessarily the result of conflict, but a realization by all stakeholders that they have common interests.

Relationship building is essential to sectoral strategies, because they require cooperation from economic development planners, employers, education and training providers, unions, community organizations and possibly other groups. Jane Addams Resource Corporation (JARC) started by building relationships with employers around a common interest: for JARC improving the income potential of local residents and for the metalworking firms increasing productivity and profits by having better-trained workers. Through the years, JARC developed relationships with foundations that allowed them to improve on the training offered. As JARC's training gained credibility, new relationships were forged with industry associations that allowed for further development of more advanced curriculum. At the same time, JARC worked with other community organizations and nonprofit agencies in workforce development to build relationships with state agencies and legislators to promote legislation that would bring new funding streams to incumbent worker training and career ladder programs. JARC is engaged in an ongoing process of building programs and building relationships to advance the programs.

In the biotechnology sectoral strategy, cooperative relationships among the staffs of state agencies, city government, and Yale University have allowed programs to complement each other. The state and city examined their respective regulations to eliminate duplication that made cumbersome paperwork for new business startups. Yale has worked with city government staff on redeveloping the business park and an abandoned building to accommodate new biotechnology businesses. This type of cooperation has made starting a new biotechnology company in the New Haven area easier than in the past.

Brownfield redevelopment is particularly complicated by the need to build political support. A comprehensive community brownfield redevelopment strategy requires the willingness to name the problem: that is, to inventory how extensive the known and potential brownfields are in a locality. This may encounter resistance by landowners as well as elected officials who perceive inventorying efforts as stigmatizing and devaluing listed properties. Again, because economic development resources are limited, a brownfield redevelopment strategy that satisfies our definition of economic development requires that the resources are not deployed in a "creaming process" for use for only the most marketable brownfields. Political pressures to show success, that most easily come with redeveloping properties in hot market areas, make it difficult to overcome this constraint. Legitimate concerns over historic patterns of environmental injustice can also pose political constraints for brownfield redevelopment. As we noted in the brownfield redevelopment chapter, because the costs of fully cleaning up a brownfield property can be so exorbitant, limited resources will largely require risk-based remediation designed for the intended future use of the brownfield property. This may not sit well with the community in which the property is located that may view restoring it to virtual greenfield status as the just solution. However, unless the parties responsible for the environmental contamination are still in business, public resources will have to be strategically employed to clean up brownfields. Often the responsible parties are no longer in business or, if they are, do not have the kinds of resources and insurance to fully clean up the contamination. Although there have been blatant examples of businesses knowingly and uncaringly contaminating properties, there are thousands of properties whose owners did not know that their business practices during the past century were causing contamination problems. Dry cleaners and gas stations are key examples found in communities everywhere.

Relationship building was a prolonged process in the battles over the PMDs discussed in the industrial retention chapter. The Washington administration was supportive of their development, but city planners had to work closely with the LIRI groups to bring other constituencies on board. The LIRI program has already built good working relationships between community organizations and employers. Real estate developers never agreed with the strategy, of course, but probably would have won had the city, employers, community groups, and labor not been in agreement on the need to preserve manufacturing.

As we noted in the commercial revitalization chapter, this strategy requires building strong community support and works best when it is pursued in conjunction with a merchant's association. Strong community support is needed because the revitalization strategy may require the rerouting of transportation routes, removing curb cuts from private commercial property, implementing more stringent zoning and design guidelines, and pruning or eliminating retail-zoned land.

The office and industrial reuse strategy also necessitates building political support to avoid redeveloping the stronger market areas to the exclusion of the weakest. For example, if limiting the use of traditional finance tools such as tax incremental financing (TIF) to the weakest market areas is needed to make this strategy equitable, political will and the support of the elected officials of city and county governments to constrain TIF use are needed.

Workforce development strategies become much more effective when good working relationships are built between economic development and workforce development offices in a city. Because the Seattle Jobs Initiative was organized as part of Seattle's Office of Economic Development, cooperation immediately became part of the organizational culture of both groups. In too many cities, the separation of these functions, combined with historical competition for resources, creates unnecessary competition. The level of cooperation among workforce development organizations in Seattle and suburban King County has resulted in organizational learning and an effective coalition for influencing state economic and workforce development policy.

Influencing the Legislative Environment

The legislative environment, as the chapter discussions illustrate, affects almost every economic development strategy to some degree. Good economic development practice often requires changing the incentives of state and local legislation to better serve the ends of promoting more equitable and sustainable development.

Sectoral strategies are affected by existing legislation to the extent that they incorporate workforce development. The amount and type of funding available for training the unemployed and incumbent workers is a factor in their effectiveness. The Seattle and King County Jobs Initiatives have been quite effective in influencing state government to structure workforce development systems and funding to advance their sectoral workforce development programs.

The brownfield redevelopment strategy may be the one that is most greatly affected by existing legislation. The brownfield problem, or barriers to redevelopment, that national policy and state and local efforts have been trying to address since the mid-1990s was in fact created by the national Comprehensive Environmental Responsibility and Liability Act (CERCLA) legislation noted in the chapter. CERCLA was intended to make "polluters pay" and to ensure competent environmental cleanup via a complex set of cleanup standards and liability assessment. However, it had the unintended consequence of making redevelopment so difficult on contaminated sites that lending institutions and private developers were unwilling to take on the higher risks and complications of the brownfields. This served to strengthen existing biases in favor of greenfield development. The evolution of brownfield policies and legislation in the second half of the 1990s has helped to clarify issues of legal liability at the national level to make brownfield redevelopment more feasible. Further, states have created their own legislation to try to stimulate brownfield redevelopment, negotiating with the EPA for authorization when such legislation intersects with national legislation. (For example, the Memorandums of Agreement that some states issue that certifies a developer has satisfied the requirements of cleanup and can expect no further action if new standards of cleanup should evolve in the future.) All local brownfield redevelopment strategies must meet federal and state regulations for cleanup. The implementation of a brownfield redevelopment strategy is easier in states that have been more proactive in creating brownfield redevelopment legislation and incentives. Likewise, it is easier in local political units that have, for example, enabled their land banks to acquire, redevelop, or both, brownfields.

An issue in the industrial retention chapter was creating more restrictive zoning legislation (planned manufacturing districts, industrial sanctuaries, and manufacturing industrial districts) that would provide greater protection for manufacturers. In Chicago, this eliminated the ability of aldermen to grant "spot" zoning allowances. In this sense, the legislation limited their political power. An ordinance had to be passed to create each of the PMDs and the same political battle preceded each one. In the cases of Portland and Seattle, state legislation required all cities to create industrial areas as a way of controlling sprawl. This type of legislation precluded any possibility of the political struggle that ensued in Chicago.

Commercial redevelopment strategies require good local legislative support in the form of authorization for overlay planning zones, changes in zoning and design standard requirements, and authorization for the use of finance tools such as TIFs. This, as we previously noted, poses a greater challenge for unincorporated inner-ring suburbs.

The reuse of office and industrial properties in central cities has been somewhat furthered at the federal level by the implementation of

Executive Order 12072 which requires federal agencies to first consider central-city properties when choosing their locations. Creating similar legislation at the state and county levels, and specifically including inner-ring suburbs, could help reinforce the reuse of central cities and inner-ring suburbs.

Like brownfield redevelopment, workforce development is largely shaped by federal legislation. The nation's series of job-training programs—the Manpower Development and Training Act, the Comprehensive Employment and Training Act, the Job Training Partnership Act, and the Workforce Investment Act—have been inadequate to meet the task of preparing entry-level workers for employment that provides living wages and advancement opportunities. Local employment and training agencies have had to rely on state funds to supplement meager federal dollars, as have community colleges. Community organizations that provide training often rely on the support of charitable foundations to provide all the services their clients need. As discussed previously, various workforce development practitioners in Seattle and surrounding King County have been very successful in shaping state-level legislation under the Workforce Investment Act and welfare reform. Examples are allowing community organizations to serve as points of entry for One-Stop Employment Centers and using TANF savings to invest in skills-upgrade training. Jane Addams Resource Corporation in Chicago, in collaboration with other community organizations and nonprofit agencies, has successfully lobbied the state to create new demonstration programs for incumbent workers with its Workforce Investment Act funds.

The Role of the Practitioner

In our Introduction to this book, we asked whether the role of economic developers was limited to tinkering at the margins rather than creating the structural change suggested by our principles. The reality is that the economic development practitioner in any city or suburb is only one of many economic agents acting within its public and private sectors. For the most part, the practitioner works on projects that have relatively small impact on the overall economy. These small projects, however, can become important catalysts for reversing a cycle of decline or otherwise improving a city's or suburb's economic development status. Further, although small, these projects move the community closer to meeting the definition of economic development offered in this book than other high-profile, expensive, subsidized projects such as a new sports stadium, a new greenfield regional mall, or subsidizing the relocation of a large manufacturing industry.

Key to whether tinkering at the margins can ultimately make a difference is the vision that informs the economic developer's work. It is for this reason in Chapter 1 that we delved in such depth into the historic role of economic developers, the de facto definitions they worked from (typically, wealth generation, business, and job creation), and then carefully articulated our alternative definition of economic development.

It is unlikely, but perhaps possible, that the high-profile projects listed previously could be implemented in a manner that satisfies one or all three principles of economic development. To do so would require a very different implementation strategy than that historically associated with such projects. In contrast, the six specific strategies we have included in this book have inherent objectives that meet one or all three of the principles. Whether or not one of the six strategies of this book is chosen, the economic development practitioner's role is to assist the community in making its choice of strategies, and in understanding how its choice of strategies furthers its vision of economic development.

References

Amekudzi, A., Fishbeck, P., Garrett, Jr., J., Kautsopoulos, H., McNeil, S., & Small, M. (1998). Techniques, cases & issues: Computer tools to facilitate brownfield development. *Public Works Management & Policy, 2*(3), 231–242.

American Economic Development Council (AEDC). (1998). *Code of ethics for the AEDC.* Rosemont, IL: Author.

American Planning Association. (1994). *Planning and community equity.* Chicago: Planners Press.

Argonne National Laboratories. (1998). *Partnering and outreach opportunities.* Chicago: Author.

Arthur Andersen LLP. (1998a). *Industrial market and strategic analysis.* Chicago: Author.

Arthur Andersen LLP. (1998b). *Market and strategic analysis for the Roosevelt and California Business Park.* Chicago: Author.

Barnes, W. R., & Ledebur, L. (1995). Local economies: The U.S. common market of local economic regions. *The Regionalist, 1,* 7–32.

Bartlett, D., & Steele, J. B. (1998, November 8). Corporate welfare. Part one. *Time,* pp. 9, 16, 23, 30, 38.

Bartsch, C. (1996, Winter). Paying for our industrial past. *Commentary,* pp. 14–23.

Bartsch, C., & Anderson, C. (1998). *Matrix of brownfield programs by state.* Northeast-Midwest Institute. Retrieved April 6, 1999, from www.nemw.org/brmatrix.html

Bartsch, C., & Collaton, E. (1995, December). *Industrial site reuse, contamination, and urban redevelopment: Coping with the challenges of brownfields.* Washington, DC: Northeast-Midwest Institute.

Bates, T. (1997). Michael Porter's conservative urban agenda will not revitalize America's inner cities: What will? *Economic Development Quarterly, 11*(1), 39–44.

Beaumont, C. E. (1997). *Better models for superstores* (Preservation Information Series). Washington, DC: National Trust for Historic Preservation.

Bernstein, J., & Mishel, L. (1995). *Trends in the low-wage labor market and welfare reform: The constraints on making work pay.* Washington, DC: Economic Policy Institute.

Betancur, J., & McCormick, L. (1985). *Industrial displacement in major cities and related policy options.* Chicago: Center for Urban Economic Development, School of Urban Planning and Policy, University of Illinois at Chicago.

Betancur, J. J., Leachman, M., Miller, A., Walker, D., & Wright, P. A. (1995). *Development without displacement. Task force background paper.* Chicago: The Chicago Rehab Network.

Biotechnology Industry Organization. (2000). *Encouraging development of the biotechnology industry: A best practices survey of state efforts.* Retrieved October 2001, from www.bio.org/govt/state_dev.html

Birch, E. L. (2001). [A rise in downtown living: A deeper look]. Unpublished data table.

Black, T. J. (1995a, June). Brownfields cleanups. *Urban Land,* pp. 47–51.

Black, T. J. (1995b). The economics of renovation in the commercial property sector. In *Reinventing real estate.* Washington, DC: Urban Land Institute.

Blank, R. M. (1997). *It takes a nation: A new agenda for fighting poverty.* Princeton, NJ: Princeton University Press.

Boston Consulting Group. (1998). *Strategies for business growth in Chicago's neighborhoods.* Chicago: Author.

Bowles, J. (1999). *Biotechnology: The industry that got away.* New York: Center for an Urban Future.

Bradshaw, T., & Blakely, E. (1999). What are "third wave" state economic development efforts? From incentives to industrial policy. *Economic Development Quarterly, 13*(3), 229–244.

Breslow, M. (2000). *Connecticut's economic development programs: High costs and inadequate job expansion.* Cambridge, MA: Commonwealth Institute, Northeast Corporate Accountability Project.

Bressi, T. (1996). The big box's final frontier. *Planning, 62*(2), 10–16.

Bucholz, D. (1998). *Good money after bad: How we waste billions on corporate incentives and why we can't stop ourselves.* Unpublished Ph.D. dissertation, Duke University.

Burstein, M. L., & Rolnick, A. J. (1994). *Congress should end the economic war among the states.* Minneapolis, MN: Federal Reserve Bank of Minneapolis.

Buss, T. (1999). The case against targeted industry strategies. *Economic Development Quarterly, 13,* 339–355.

Cagan, J., & deMause, N. (1998). *Field of schemes: How the great stadium swindle turns public money into private profit.* Monroe, ME: Common Courage Press.

California Trade and Commerce Agency, Office of Economic Research. (2000). *Biotechnology in California.* Sacramento, CA: Author.

Campbell, S. (1996). Green cities, growth cities, just cities? Urban planning and the contradictions of sustainable development. *Journal of the American Planning Association, 62*(3), 296–312.

Carlson, V., & Wiewel, W. (1991, Fall). Strategic planning in Chicago. *Economic Development Commentary,* pp. 17–22.

Castells, M. (1996). *The rise of the network society.* Malden, MA: Blackwell.

Chanen, J. S. (1997, July 27). In Chicago's loop, conversions to condos. *New York Times,* Sect. 9, p. 5.

Chicago Association of Neighborhood Development Organizations. (1988). *Land use techniques for industrial preservation: Report to the National Trust for Historic Preservation.* Chicago: Author.

Christensen, M. F., & Ackerman, K. B. (1996). Warehouse distribution facilities: Emerging industry trends and future market implications. In *Creating tomorrow's competitive advantage* (pp. 18–27). Washington, DC: Urban Land Institute.

City of Chicago, Department of Economic Development, Department of Planning. (1988). *The Clybourn Corridor planned manufacturing district.* Chicago: Author.

City of Emeryville. (1998, November). *Project status report, Emeryville brownfields pilot project.* Emeryville, CA: Author.

City of Kalamazoo. (n. d.). *Kalamazoo brownfield redevelopment initiative.* Kalamazoo, MI: Author.

City of Seattle, Office of Economic Development. (1994). *Strategic plan of the Seattle Office of Economic Development.* Seattle, WA: Author.

City of Seattle, Office of Economic Development. (1997). Brochure. Seattle, WA: Author.

City of Seattle, Office of Economic Development. (1999). *SJI strategic plan.* Seattle, WA: Author.

City of Seattle, Office of Management and Planning. (1997). *Survey of industrial land retention strategies in selected North American cities.* Seattle, WA: Author.

Clark, P., & Dawson, S. (1995). *Jobs and the urban poor: Privately initiated sectoral strategies.* Washington, DC: Aspen Institute.

Clavel, P. (1986). *The progressive city: Planning and participation, 1969–1984.* New Brunswick, NJ: Rutgers University Press.

Clavel, P., & Wiewel, W. (Eds.). (1991). *Harold Washington and the neighborhoods: Progressive city government in Chicago, 1983–1987.* New Brunswick, NJ: Rutgers University Press.

Coffee, H. E. (1994, February). Location factors: Business as usual, more or less. *Site Selection,* pp. 32–33.

Colman, G. J. (1992, August). Meet me at the mall. *Children's Business.* Retrieved November 23, 1998, http://web.lexis-nexis.com/universe

Congressional Research Service. (1995). *Summaries of environmental laws administered by the Environmental Protection Agency.* Washington, DC: Government Printing Office.

Connecticut Department of Economic and Community Development. (2000). *Connecticut town profiles.* Hartford, CT: Author.

Connecticut's economic competitiveness strategy. (1998). Retrieved November 6, 2001, from www.cerc.com/cerc/cerc.nsf/pages/clusters.

Conway, M. (1999). *The garment industry development corporation: A case study of a sectoral employment development approach.* Washington, DC: Aspen Institute.

Davis, T. S., & Margolis, K. D. (1997). *Brownfields: A comprehensive guide to redeveloping contaminated property.* Chicago: American Bar Association.

Denver stands out in mini-trend toward downtown living. (1998, December 29). *New York Times,* p. 10.1.

Design Center for American Urban Landscape. (1999). *Reframing the 1945–1965 suburb: A national conference on contemporary public policy, and scholarship.* Conference Proceedings, University of Minnesota, January 21–23, 1999. Retrieved March 15, 1999, from www1.umn.edu/dcaul

DeVries, J. (1998, June n. d.). Parking lessons: From brownfields to industrial parks, the key is financing. *Brownfield News,* n. p.

DeVries, J., & Lenz, G. (1999, June). Recentralization. *Urban Land,* pp. 68–69, 73, 94.

Dewar, M. (2000). *The Detroit empowerment zone's effect on economic opportunity: Employers' responses to the zone's programs and incentives.* Ann Arbor, MI: Urban and Regional Planning Program at the University of Michigan.

Dinsmore, C. (1996, June). State initiatives on brownfields. *Urban Land,* pp. 37–42.

Dostaler, M. A. (2000, September/October). Finding room to grow. *Connecticut Business,* pp. 76–79.

Dresser, L., & Rogers, J. (1998). Networks, sectors, and workforce learning. In R. Giloth (Ed.), *Jobs and economic development* (pp. 64–84). Thousand Oaks, CA: Sage.

Ducharme, D. (1991). How a community initiative became city policy. In P. Clavel & W. Wiewel (Eds.), *Neighborhood and economic development policy in Chicago 1983–1987* (pp. 221–237). New Brunswick, NJ: Rutgers University Press.

Ducharme, D., Giloth, R., & McCormick, L. (1986). *Business loss or balanced growth: Industrial displacement in Chicago.* Chicago: City of Chicago Department of Economic Development.

Dunn, B. C., & Steinemann, A. (1998). Industrial ecology for sustainable communities. *Journal of Environmental Planning and Management,* 4(6), 661–672.

Dwortzan, M. (1998). The greening of industrial parks. *MIT's Technology Review*, *100*(9), 18–19.

Edin, K., & Lein, L. (1997). *Making ends meet: How single mothers survive welfare and low-wage work*. New York: Russell Sage.

Eisinger, P. K. (1988). *The rise of the entrepreneurial state*. Madison: University of Wisconsin Press.

Eisinger, P. (1995). State economic development in the 1990s: Politics and policy learning. *Economic Development Quarterly*, *9*(2), 146–158.

Ehrenfeld, J., & Gertler, N. (1997). Industrial ecology in practice: The evolution of interdependence at Kalundborg. *Journal of Industrial Ecology*, *1*(1), 67–79.

Erb, G. (1999, October 5). Suburban cities investing in economic development in Puget Sound. *Business Journal*, *1*(12), n. p.

Fainstein, S., & Gray, M. (1995). Economic development strategies for the inner city: The need for governmental Intervention. *Review of Black Political Economy*, *24*(Special Issue: Responses to Michael Porter's model of inner-city redevelopment), 29–38.

Ferguson, B. W., Miller, M. M., & Liston, C. (1996, Winter). Retail revitalization. *Commentary*, pp. 4–13.

Fingeret, L. (1998, February 16). Chicago approves 1,500-acre TIF chunk. *Real Estate Journal*, p. 14.

Finkle, J. A. (1999). The case against targeting might have been more . . . targeted. *Economic Development Quarterly*, *13*(4), 361–364.

Fitzgerald, J. (1998). Is networking always the answer? Networking among community colleges to increase their capacity in business outreach. *Economic Development Quarterly*, *12*, 30–40.

Fitzgerald, J. (1999). *Principles and practices for creating systems reform in urban workforce development*. Discussion paper commissioned by the Brookings Institution for the Annie E. Casey Foundation Jobs Initiative Policy Retreat. Chicago: Great Cities Institute, University of Illinois at Chicago.

Fitzgerald, J., & Carlson, V. (2000). Ladders to a better life. *American Prospect*, *11*(15), 54–60.

Fitzgerald, J., & Cox, K. (1990). Urban economic development strategies in the USA. *Local Economy*, *4*, 278–289.

Fitzgerald, J., & Jenkins, D. (1997). *Making connections: Best practice efforts by U.S. community colleges to connect the urban poor to education and employment*. Report prepared under contract to the Annie E. Casey Foundation. Chicago: Great Cities Institute, University of Illinois at Chicago.

Fitzgerald, J., & McGregor, A. (1993). Labor-community initiatives in worker training in the United States and United Kingdom. *Economic Development Quarterly*, *7*(2), 172–182.

Fitzgerald, J., & Meyer, P. B. (1986). Recognizing constraints to local economic development. *Journal of the Community Development Society, 17*(2), 115–126.

Fitzgerald, J., & Patton, W. (1994). Race, job training, and economic development: Barriers to racial equity in program planning. *Review of Black Political Economy, 23*(2), 93–112.

Fitzgerald, J., Putterman, J., & Theodore, N. (2000). *Increasing job retention and advancement among TANF clients: Recommendations to Illinois Department of Human Services.* Chicago: University of Illinois at Chicago, Great Cities Institute.

Fitzgerald, J., & Rasheed, J. (1998). Salvaging an evaluation from the swampy lowlands. *Evaluation and Program Planning, 21*(2), 199–209.

Fitzgerald, J., & Sutton, S. (2000). *One stops: Something for everyone?* Washington, DC: U.S. Department of Labor.

Fort, R. (1997). Direct democracy and the stadium mess. In Roger G. Noll & Andrew Zimbalist (Eds.), *Sport, jobs and taxes* (pp. 146–177). Washington, DC: Brookings Institution.

Foster-Bey, J. (1997). Bridging communicites: Making the link between regional economies and local community economic development. *Stanford Law Review, 8*, n. p.

Gann, D. (1992). *Intelligent buildings: Producers and users.* University of Sussex: Science Policy Research Unit.

Garland, S., & Galuszka, P. (1997, June 2). The burbs fight back. *Business Week*, pp. 147–148.

Geddes, R. (1997). Metropolis unbound. *American Prospect, 8*(35), 40–46.

Gentry, C. R. (2000, September 1). Reinvent strip centers. *Chain Store Age.* Retrieved October 5, 2001, from web.lexis-nexis.com/universe/docum. . . 1A1&_md5=06408ee877bacf708f9bf59fcbd6942cb

Gibbs, R. (2000, April 13). Lecture presented to "Intown Retail Trends Seminar," Atlanta Development Authority, Atlanta, GA.

Giloth, R. (1998). Jobs and economic development. In R. Giloth (Ed.), *Jobs and economic development* (pp. 1–18). Thousand Oaks, CA: Sage.

Giloth, R., & Betancur, J. (1988, Summer). Where downtown meets the neighborhood: Industrial displacement in Chicago, 1978–1987. *APA Journal*, pp. 279–290.

Goodman, R. (1979). *The last entrepreneurs.* New York: Simon & Schuster.

Goozner, M. (1998). The porter prescription. *American Prospect, 9*(38), 20–24.

Graham, S., & Marvin, S. (1995). *Telecommunications and the city.* London: Routledge.

Granovetter, M. (1985, November). Economic action and social structure: The problem of embeddedness. *American Journal of Sociology*, 481–510.

Greening Wal-Mart. (1993, March). *Interiors*, pp. 63–64.

Greenstein, R., & Wiewel, W. (2000). *Urban suburban interdependencies.* Cambridge, MA: Lincoln Institute of Land Policy.

Harrison, B., & Bluestone, B. (1988). *The great u-turn: Corporate restructuring and the polarizing of America.* New York: Basic Books.

Harrison, B., & Glasmeier, A. (1997, February). Why business alone won't re-develop the inner city: A friendly critique of Michael Porter's approach to urban revitalization. *Economic Development Quarterly, 11*(1), 28–38.

Harrison, B., & Kanter, S. (1978). The political economy of state job-creation business incentives. *Journal of the American Institute of Planners, 44*(2), 424–435.

Harrison, B., & Weiss, M. (1998). *Workforce development networks.* Thousand Oaks, CA: Sage.

Harrison, B., Weiss, M., & Gant, J. (1998). *Building bridges: Community development corporations and the world of employment training.* New York: Ford Foundation.

Hazel, D. (2000a, August 1). Strip center marketing looks beyond newspapers. *Shopping Centers Today.* Retrieved December 11, 2000, from www.icsc.org/srch/sct/current/sct0800/02.html

Hazel, D. (2000b, November 1). Tenants become landlords. *Shopping Centers Today.* Retrieved December 11, 2000, from www.icsc.org/srch/sct/current/sct1100/cover.html

Hershey, A. M., & Pavetti, L. (1997). Turning job finders into job keepers. *The Future of Children, 7*(1), 74–86.

Hill, E. W., Wolman, H. L., & Ford, III, C. C. (1995). Can suburbs survive without their central cities? Examining the suburban dependence hypothesis. *Urban Affairs Review, 31*(2), 147–174.

Hinkley, S., Hsu, F., LeRoy, G., & Tallman, K. (2000). *Minding the candy store: State audits of economic development.* Washington, DC: Good Jobs First of the Institute on Taxation and Economic Policy.

Hinz, G. (1997, July). The city that TIFs. In *Crain's Chicago Business* (pp. 1, 11–15). Chicago: Crain Communications.

Hinz, G. (2000, March 6). The new downtown. In *Crain's Chicago Business* (pp. 5–6). Chicago: Crain Communications.

Hird, J. A. (1994). *Superfund: The political economy of environmental risk.* London: The Johns Hopkins University Press.

HOH Associates. (1993, May 25). *Sandy Springs revitalization plan.* Atlanta, GA: Author.

Holzer, H. (1996). *What employers want: Job prospects for less-educated workers.* New York: Russell Sage.

Hughes, M. A. (2000). Federal roadblocks to regional cooperation: The administrative geography of federal programs in larger metropolitan areas. In R. Greenstein & W. Wiewel (Eds.), *Urban-suburban interdependencies* (pp. 161–180). Cambridge, MA: Lincoln Institute of Land Policy.

Iannone, D. (1996). Increasing public and private capital to brownfields, or how shall we pay for the sins of the past. *Infrastructure, 1*(4), 18–23.

Illinois Department of Employment Security. (1996). *Where workers work.* Springfield, IL: Author.

International Council of Shopping Centers (ICSC). (2000). Scope 2000. *Scope USA.* Retrieved December 11, 2000, from www.icsc.org/srch/rsrch/scope/current/index.html

Jacobs, J. (1979). *Bidding for business: Corporate auctions and the fifty disunited states.* Washington, DC: Public Interest Research Group.

Jenkins, D. (1999). *Beyond welfare-to-work: Bridging the low-wage-livable-wage employment gap* (Working paper). Chicago: University of Illinois at Chicago, Great Cities Institute.

Johnson, G. (1996, October 29). Big-box boom; Superstores sprout as O.C. strip malls wither. *Los Angeles Times,* p. 9.1.

Jossi, F. (1998). Rewrapping the big box. *Planning, 64*(8), 16–19.

Kerch, S. (1997, July 6). Loft conversion binds generations to urban housing. *Chicago Tribune,* p. 1.

King, J. (1988). Protecting manufacturing from yuppies and other invaders. *The Best of Planning,* n. p.

Kleiman, N. S. (2000). *The sector solution.* New York: Center for an Urban Future.

Klemanski, J. S. (1989). Tax increment financing: Public funding for private economic development projects. *Policy Studies Journal, 17*(2), 656–671.

Krumholz, N. (1991). Equity and local economic development. *Economic Development Quarterly, 5*(4), 291–300.

Krumholz, N., & Clavel, P. (1994). *Reinventing cities: Equity planners tell their stories.* Philadelphia: Temple University Press.

Krumholz, N., & Forester, J. (1990). *Making equity planning work.* Philadelphia: Temple University Press.

Leigh, N. G. (1994). Focus: Environmental constraints to brownfield redevelopment. *Economic Development Quarterly, 8*(4), 325–328.

Leigh, N. G. (1996). Fixed structures in transition: The changing demand for office and industrial infrastructure. In Daniel Knudsen (Ed.), *The transition to flexibility* (pp. 137–154). Norwell, MA: Kluwer Press.

Leigh, N. G. (2000). *Promoting more equitable brownfield redevelopment: Promising approaches for land banks and other community land development entities.* Boston: Lincoln Institute on Land Policy.

Leigh, N. G., & Coffin, S. L. (2000). Industrial era's brownfield legacy: Characterizing the redevelopment challenge. *Journal of Urban Technology, 7*(3), 1–18.

Leigh, N. G., & Hise, R. (1997). *Community brownfield guidebook: Assessing and resolving environmental barriers to redevelopment.* Atlanta: Georgia Tech Research Corporation.

Leigh, N. G. (1994). *Stemming middle class decline: The challenge to economic development planning.* New Brunswick, NJ: Rutgers University, CUPR Press.

Leigh, N. G. (1999). *The influence of new flexibility and technology requirements on central city land use devoted to office and industrial property* (Working paper) Cambridge, MA: Lincoln Institute of Land Policy.

Leigh, N. G., & Realff, M. (2001). *A framework for geographically sensitive and efficient recycling networks.* Manuscript submitted for publication.

Leigh-Preston, N. (1985, July). *Industrial transformation, economic development, and regional planning.* Chicago: Council of Planning Librarians Bibliography 154.

LeRoy, G. (with Healey, R., Doherty, D., & Kerson, R.). (1994). *No more candy store.* Chicago: Federation for Industrial Retention and Renewal.

Levine, J. (1994, December). Message posted to planet@listserv.acsu.-buffalo.edu

Lidman, R. M. (1995). *The family income study and Washington's welfare population: A comprehensive review.* Olympia: Washington State Institute for Public Policy.

Louisville/Jefferson County Landbank Authority. (1997). Resolution 53.

Loukaitou-Sideris, A. (2000). Revisiting inner-city strips: A framework for community and economic development. *Economic Development Quarterly, 14*(2), 165–181. Retrieved January 29, 2001, from http://ehostvgw6.epnet.com/print

Luger, M., & Goldstein, H. (1993). Research parks as public investments: A critical assessment. In R. Bingham, E. Hill, & S. White (Eds.), *Financing economic development* (pp. 142–159). Newbury Park, CA: Sage.

MacFarlane, R., Belk, J., & Clark, J. A. (1994). Confronting the impacts of environmental strict liability on brownfield redevelopment. *Environmental Claims Journal, 6*(4), 459–479.

Mahtesian, C. (1994, November). Romancing the smokestack. *Governing,* pp. 36–40.

Mainstreet. (1999). *Marketing an image for main street.* Washington, DC: National Trust for Historic Preservation.

Marcelli, E., Baru, S., & Cohen, D. (2000). *Planning for shared prosperity or growing inequality? An in-depth look at San Diego's leading industry clusters.* San Diego, CA: Center on Policy Initiatives.

Marris, P. (1982). *Meaning and action: Community planning and conceptions of change.* London: Routledge & Kegan Paul.

Massachusetts Annotated Laws, ch. 29, sec. 2W, 1999. *Brownsfield revitalization fund.* Retrieved April 16, 1999, from www.magnet.state.ma.us/mdfa/predev.htm

Massey, D. (1984). *Spatial divisions of labor: Social structures and the geography of production.* New York: Methuen.

Massey, D. S., & Denton, N. (1993). *American apartheid: Segregation and the making of the underclass.* Cambridge, MA: Harvard University Press.

Massing, M. (2000). Making work pay: Knowing what we do now, how should Congress change welfare reform? *American Prospect, 11*(15), 30–39.

McCartney, J. (2000, February 1). Big-box strategy: Emulating the little guy. *Shopping Centers Today.* Retrieved December 11, 2000, from www.icsc.org/srch/sct/current/sct0200/05.html

McIntosh, W., & Whitaker, W. (1998, January/February). What REITs mean to you. *Corporate Real Estate Executive,* pp. 24–27.

Metro-Chicago Office Guide, Second Quarter. (1997). Chicago: Law Bulleting.

Metzger, J. (1996). The theory and practice of equity planning: An annotated bibliography. *Journal of Planning Literature, 11*(1), 112–126.

Meyerson, H. (2001). California's progressive mosaic. *American Prospect, 12*(11), 12–13.

Mier, R. (1993). *Social justice and local development policy.* Newbury Park, CA: Sage.

Mier, R. (1994). Some observations on race in planning. *Journal of the American Planning Association, 2*(60), 235–240.

Mier, R., & Fitzgerald, J. (1991). Managing local economic development. *Economic Development Quarterly, 5*(3), 268–279.

Mier, R., & Giloth, R. (1983). Hispanic employment opportunities: A case of internal labor markets and weak tied social networks. *Social Science Quarterly, 66*(2), 296–309.

Mier, R., & Giloth, R. (1993). Cooperative leadership for community problem solving. In R. Mier (Ed.), *Social justice and local development policy* (pp. 165–181). Newbury Park, CA: Sage.

Mier, R., & Moe, K. J. (1991). Decentralized development: From theory to practice. In P. Clavel & W. Wiewel (Eds.), *Harold Washington and the neighborhoods: Progressive city government in Chicago, 1983–1987* (pp. 64–99). New Brunswick, NJ: Rutgers University Press.

Mier, R., Moe, K., & Sherr, I. (1986). Strategic planning and the pursuit of reform, economic development, & equity. *Journal of the American Planning Association, 52*(3), 299–309.

Milner, R. R. (1997, Summer). Logistics: Driving new patterns in industrial real estate. *Professional Report,* n. p.

Modern traffic handling capability in elevator retrofit. (1986, June). *Buildings,* p. 158.

Mollenkopf, J. (1983). *The contested city.* Princeton, NJ: Princeton University Press.

Molotch, H. (1976). The city as a growth machine: Toward political economy of place. *American Journal of Sociology, 82*, 309–332.

Moss, P., & Tilly, C. (1996). *Informal hiring practices, racial exclusion, and public policy*. Lowell: University of Massachusetts at Lowell.

Mt. Auburn Associates. (1995). *Sectoral targeting: A tool for strengthening state and local economies*. Washington, DC: Council of State Community Affairs Agencies.

Mumford, S. (1993, November). Turning on savings with lighting retrofits. *Buildings*, pp. 40–42.

Muschamp, H. (1997, October 19). Becoming unstuck on the suburbs. *New York Times*, pp. 4.4.

Myers, J. (1994, April). Fundamentals of production that influence industrial facility designs. *Appraisal Journal, 62*(2), 296–302.

National Trust for Historic Preservation. (2000). *Mainstreet catalog*. Washington, DC: Author. Material reprinted with permission from the National Trust Forum, The National Trust for Historic Preservation, 1785 Massachusetts Avenue NW, Washington, DC 20036; www.nationaltrust.org.

Neckerman, K., & Kirschenman, J. (1991). Hiring strategies, racial bias, and inner-city workers. *Social Problems, 38*(3), 433–447.

New Jersey Office of State Planning. (1995, December). *Big box retail* [Memo], *1*(2).

Nexus Associates. (1999). *A record of achievement: The economic impact of the Ben Franklin partnership*. Belmont, MA: Author.

Noonan, F., & Vidich, C. (1992). Decision analysis for utilizing hazardous waste site assessments in real estate acquisition. *Risk Analysis, 12*(2), 245–251.

Norman, A. (1999). *Slam-dunking Wal-Mart!* Atlantic City, NJ: Raphel Marketing.

Nunnink, K. K. (Ed.). (1994). *Viewpoint 1994*. Minneapolis, MN: Valuation International.

Off the urban rust heap, a factory goes to work. (1999, January 10). *New York Times*, p. 3.1.

Okagaki, A., Palmer, K., & Mayer, N. S. (1998). *Strengthening rural economies: Programs that target promising sectors of local economy*. Washington, DC: Center for Community Change.

Orfield, M. (1997). *Metropolitics: A regional agenda for community and stability*. Washington, DC: Brookings Institution.

Orfield, M. (1998, November 15). Salvaging suburbia; How to stop communities from growing farther and farther apart. *Los Angeles Times*, Part M, p. 1.

Osborne, D., & Gaebler, T. (1993). *Reinventing government: How the entrepreneurial spirit is transforming the public sector*. New York: Plume.

Osterman, P. (1988). *Employment futures*. New York: Oxford University Press.

Osterman, P. (1993). Why don't "they" work? Employment patterns in a high pressure economy. *Social Science Research, 22*(2), 115–130.

Patchin, P. (1994). Contaminated properties and the sales comparison approach. *Appraisal Journal, 62*(3), 402–409.

Patchin, P. J. (1988). Valuation of contaminated properties. *Appraisal Journal, 56*(1), 7–16.

Patchin, P. J. (1991). Contaminated properties: Stigma revisited. *Appraisal Journal, 59*(2), 167–178.

Pavetti, L. (1997). *Against the odds: Steady employment among low-skilled women. A report to the Annie E. Casey Foundation.* Washington, DC: The Urban Institute.

Pavetti, L., & Acs, G. (1996). *Moving up, moving out or going nowhere? A study of the employment patterns of young mothers and the implications for welfare reform.* Washington, DC: Urban Institute.

Peck, J. (1996). *Work place: The social regulation of labor markets.* New York: Guilford.

Peirce, N. (1996). Big box retailers: Time to ask for performance bonds? *Nation's Cities Weekly, 19*(2), n. p.

Pepper, E. (1997). *Lessons from the field.* Washington, DC: The League.

Percival, R. V. (2000). *Environmental regulation: Law, science, and policy* (3rd ed.). Boston: Aspen Law & Business.

Persky, J., & Wiewel, W. (2000). The distribution of costs and benefits due to employment deconcentration. In R. Greenstein & W. Wiewel (Eds.), *Urban-suburban interdependencies* (pp. 49–70). Cambridge, MA: Lincoln Institute of Land Policy.

Podmolik, M. E. (1997, February 25). Tripp plans to take over Spiegel site in Bridgeport. *Chicago Sun-Times,* n. p.

Point 10 of the "Code of Ethics" for the AEDC. (1998). Rosemont, IL: American Economic Development Council.

Porter, M. (1997). New strategies for inner-city economic development. *Economic Development Quarterly, 11*(1), 11–27.

Portney, P., & Probst, K. (1994, Winter). Cleaning up superfund. *Resources,* pp. 2–5.

Pouncy, H., & Mincy, R. B. (1995). Out of welfare: Strategies for welfare-bound youth. In D. Smith Nightingale & R. H. Haverman (Eds.), *The work alternative: Welfare reform and the realities of the job market* (n. p.). Washington, DC: Urban Institute Press.

Powell, J. A. (2000). Addressing regional dilemmas for minority communicites. In Bruce Katz (Ed.), *Reflections on regionalism* (pp. 218–246). Washington, DC: Brookings Institution.

President's Council on Sustainable Development. (1999). *Eco-industrial park workshop proceedings.* Retrieved November 18, 2001, from http://clinton2.nara.gov/PCSD/Publications/Eco_Workshop.html

Reinbach, A. (1993a, February). The green retrofit. *Buildings*, pp. 32–35.

Reinbach, A. (1993b, November). Wire/cable distribution: More than meets the eye. *Buildings,* pp. 48–54.

Rosenberg, R. (2000, March 12). A boost from biotech: Yale backs start-ups to help make ailing New Haven a major biomedical center. *Boston Globe*, p. G1.

Ross, C. L., & Leigh, N. G. (2000, February). Planning, urban revitalization and the inner city: An exploration of structural racism. *Journal of Planning Literature, 14*(3), n. p.

Rossi, F. (1998, August). Rewrapping the big box. *Planning*, pp. 16–18.

Rubin, H. (1988). Shoot anything that flies; Claim anything that falls. *Economic Development Quarterly, 2*(2), 236–251.

Rusk, D. (1993). *Cities without suburbs.* Washington, DC: Woodrow Wilson Center Press.

Rusk, D. (2000). Growth management: The core regional issue. In Bruce Katz (Ed.), *Reflections on regionalism* (pp. 78–106). Washington, DC: Brookings Institution.

San Diego Association of Governments. (1998). *Creating prosperity for the San Diego region.* San Diego, CA: Author.

Sandy Springs Revitalization, Inc. (SSRI). (1997, October). *Sandy Springs framework plan.* Atlanta, GA: Author.

Savitch, H. V., Collins, D., Sanders, D., & Markham, J. P. (1993). Ties that bind: Central cities, suburbs and the new metropolitan region. *Economic Development Quarterly, 7*(4), 341–357.

Saxenian, A. L. (1985). Review 3: Let them eat chips. *Society and Space, 3*, 121–127.

Schneider, K. (1999, Winter). Cleveland suburbs blaze trail, work together to stop sprawl. *Great Lakes Bulletin, 4*(1), n. p.

Schwartz, E. (1994). Reviving community development. *American Prospect, 5*(19). Retrieved November 18, 2001, from www.prospect.org/print/V5/19/schwartz-e.html

Seattle Jobs Initiative. (2001). *Seattle jobs initiative employment tracking database.* Seattle, WA: Author.

Silver, J. (1997, December 5–11). Rehabilitation propels building into cyber era. *Atlanta Business Chronicle,* n. p.

Simmons, L. B. (1994). *Organizing in hard times: Labor and neighborhoods in Hartford.* Philadelphia: Temple University Press.

Simons, R. A., & Iannone, D. T. (1997, June). Brownfields: Supply and demand analysis. *Urban Land*, pp. 36–39.

Soil washing proving a good bet. (1993, July 19). *Environmental News Reporter,* p. 27.

Solo, J. A. (1995, Spring). Urban decay and the role of superfund: Legal barriers to redevelopment and prospects for change. *Buffalo Law Review, 43,* 285–327.

Stanback, T. M. Jr., (1991). *The new suburbanization challenge to the central city*. San Francisco: Westview Press.

Stanback, T. M., Jr., & Knight, R. (1976). *Suburbanization and the city*. Montclair, NJ: Allanheld, Osmun & Co.

Stillman, A. (1999). *Better models for chain drugstores*. Washington, DC: National Trust.

Storper, M., & Walker, R. (1989). *The capitalist imperative: Territory, technology and industrial growth*. New York: Basil Blackwell.

Strawn, J., & Martinson, N. (1998). *Taking the next steps: How welfare reform and workforce development can promote steady work and better jobs*. Washington DC: Center for Law and Social Policy.

Taylor, D., & Smalling Archer, J. (1996). *Up against the Wal-Marts: How your business can prosper in the shadow of the retail giants*. New York: AMACOM.

Theodore, N. (1995). *When the job doesn't pay: Contingent workers in the Chicago metropolitan area*. Chicago: Chicago Urban League.

Theodore, N., & Carlson, V. (1998). Targeting job opportunities: Developing measures of local employment. *Economic Development Quarterly, 12*(2), 137–149.

U.S. Conference of Mayors, Brownfields Task Force. (2000, February). *Recycling America's land: A national report on brownfield redevelopment*. Washington, DC: Author.

U.S. Congress, Office of Technology Assessment. (1995). *State of the states on brownfields: Programs for cleanup and reuse of contaminated sites*. Washington, DC: Government Printing Office.

U.S. Department of Housing and Urban Development. (1999a). *New markets: The untapped retail buying power in America's inner cities*. Washington, DC: Office of Policy Development and Research.

U.S. Department of Housing and Urban Development. (1999b). *Now is the time: Places left behind in the new economy*. Washington, DC: Author.

U.S. Department of Housing and Urban Development. (1999c). *The state of the cities, 1999*. Washington, DC: Author.

U.S. Department of Housing and Urban Development. (2000). *The state of the cities, 2000: Megaforces shaping the future of the nation's cities*. Washington, DC: Author.

U.S. Department of Labor, Bureau of Labor Statistics. (2000). *Chicago-Gary-Kenosha, IL-IN-WI national compensation survey, October, 1999* (Bulletin 3100–76). Washington, DC: Author.

U.S. Department of Labor, Bureau of Labor Statistics (1990, 1992, 1994, 1996, 1998, 2000a). *State and area employment reports*. Washington, DC: Author.

U.S. Department of Labor, Bureau of Labor Statistics. (1990, 1992, 1994, 1996, 1998, 2000b). *National employment, hours and earnings*. Washington, DC: Author.

U.S. Environmental Protection Agency. (1995, January). *The brownfields action agenda*. Washington, DC: Author.

U.S. Environmental Protection Agency. (1996). *Integrated approaches to brownfield redevelopment: Smart growth network*. Washington, DC: Author.

Vaugh, R. (1979). *State taxation and economic development*. Washington, DC: Council of State Planning.

Vidal, A. (1992). *Rebuilding communities: A national study of urban community development corporations*. New York: Community Development Research Center.

Waits, M. J., & Howard, G. (1996). Industry clusters. *Commentary, 20*, 5–11.

Walker, R. (1984). Industrial location policy: False conclusions. *Built Environment, 6*, 105–114.

Walton, J. (1979). A systematic survey of community power research. In M. Aiken & P. Mott (Eds.), *The structure of community power* (pp. 443–464). New York: Random House.

Weber, J., Gogoi, P., Palmer, A. T., & Crockett, R. (2000, October 16). Chicago blues. *Business Week*, pp. 163–172.

Weber, R. (1999, Summer). *Making tax increment financing work for workforce development*. Chicago: University of Illinois at Chicago, Great Cities Institute.

Wiewel, W. (1999). Policy research in an imperfect world: Response to Terry F. Buss, the case against targeted industry strategies. *Economic Development Quarterly, 13*, 357–360.

Wiewel, W., Persky, J., & Felsenstein, D. (1994, November). Are subsidies worth it? How to calculate the costs and benefits of business incentives. *Government Finance Review*, n. p.

Wiewel, W., & Siegel, W. (1990). Industry task forces as pragmatic planning: The effect of ideology, planning process, and economic context on strategy selection. *Journal of Planning Education and Research, 9*, 107–115.

Wilson, Marianne. (1993, July). Wal-Mart makes a green statement. *Shopping Center Age, 69*(7), Sect. 1, p. 23.

Wisconsin Department of Natural Resources. (n. d.). *Summary of Wisconsin's new brownfields initiatives and programs*. Madison: Author.

Wright, E. O., & Dwyer, R. (1999, July). *The American jobs machine: The trajectory of good and bad jobs, 1983–1997*. Paper presented for the annual meeting of the Society for the Advancement of Socioeconomics, Madison, WI.

Index

Retrofitting, of office properties, 190-193
Reuse of structures:
    "Big Box" unused facilities, 146
    brownfield property, 176, 229
    historical buildings, 145
    industrial property, 36-37, 175-186,
        228, 229
    legislation and, 238-239
    "live and work" space, 179
    office properties, 36-37, 164-175, 228
    relationship building and, 237
Rice, Mayor Norman, 199, 200
Rice, William B., 60
Robbins (IL), 18-20
Rookery Building (Chicago), 169
Ryan, Mary Jean, 200

SACED. See State Agenda for Community
    Economic Development
Salt Lake City, downtown population
    statistics, 172
San Antonio (TX), downtown population
    statistics, 172
San Diego:
    downtown population statistics, 172
    trade-offs in economic
        development, 234
San Diego Association of Governments
    (SANDAG), 43
San Francisco, downtown population
    statistics, 172
Sandy Springs (GA), commercial
    revitalization, 150, 151-160, 232
Sandy Springs Business Association
    (SSBA), 157-158
Sandy Springs Revitalization, Inc. (SSRI),
    152-160
Science Park (New Haven), 58-59, 66
Sears, relocation out of Chicago, 167
Seattle:
    brownfield redevelopment, 215
    downtown population statistics, 172
    industrial retention, 124-125, 129
    job-centered economic
        development, 196
    manufacturing employment in, 104
    workforce development, 196, 199-213,
        215-222, 228, 239
Seattle Jobs Initiative (SJI), 196, 199-213,
    215-222

Sectoral Employment Development
    Learning Project (SEDLP), 48-49
Sectoral strategies, 35, 39-68
    biotechnology in Connecticut, 53-63,
        65-67, 226, 229, 235
    defined, 44
    effectiveness of, 47-49
    goals of, 44
    government-led, 45
    Jane Addams Resource Corporation,
        42, 49-53, 63-67, 226, 229,
        230, 235, 239
    legislation and, 237
    planning process for, 46-47
    relationship building and, 235
    reuse of industrial space, 229
    suburbs, 231
    targeted industry strategies, 40, 45-49,
        201-202, 219
    trade-offs, 233-234
    Type 1, 40, 42, 64-65
    Type 2, 40, 43, 44-45
    typology of, 40-43
Shopping centers, 132-133, 135
    "Big Box" retailing, 133, 140-147, 161
    strip shopping centers, 147-151
Shreveport (LA), downtown population
    statistics, 172
SJI. See Seattle Jobs Initiative
Small-quantity generators, brownfields,
    79-80
Smart buildings, 170-171, 173-175,
    190-193
SMOA. See Superfund Memorandum of
    Agreement
Social goals, economic development
    and, 11
Social justice, economic development and,
    17-21, 29-30, 33
"Soil washing," brownfield
    redevelopment, 85
Sonnhalter, James, 126
Springfield (MA), Machine Action Project
    (MAP), 42
SSBA. See Sandy Springs Business
    Association
SSRI. See Sandy Springs
    Revitalization, Inc.
St. Louis, downtown population
    statistics, 172

About the Authors

Joan Fitzgerald is Associate Professor and Associate Director of the Center for Urban and Regional Policy at Northeastern University. Prior to moving to Boston in 2000, she was on the faculty at the College of Urban Planning and Public Affairs at the University of Illinois at Chicago for nine years and a faculty fellow at the Great Cities Institute for three years. Fitzgerald has taught urban economic development and related courses for 12 years. Her research focuses on linking workforce and economic development. She has published in *Economic Development Quarterly, The Review of Black Political Economy, Urban Affairs Quarterly, Urban Education,* and *The American Prospect.* She is currently working on a book, *Moving Up in the New Economy.* The book focuses on how to build career ladders to living-wage jobs into workforce development programs. Fitzgerald holds B.A., M.A., and Ph.D. degrees from the Pennsylvania State University.

Nancey Green Leigh is Associate Professor, specializing in economic development planning, in the Department of City and Regional Planning at the Georgia Institute of Technology. She obtained her B.A. in urban studies and a master's in regional planning from the University of North Carolina at Chapel Hill, and a master's in economics and a Ph.D. in city and regional planning from the University of California at Berkeley. She is a former Woodrow Wilson Fellow and Regents Fellow of the University of California at Berkeley and past Vice President of the Association of the Collegiate Schools of Planning. She is a member of the American Institute of Certified Planners. Leigh teaches, conducts research, and publishes in the areas of local economic development planning, urban and regional development, industrial restructuring, and brownfield redevelopment. She is the author of *Stemming Middle Class Decline: The Challenge to Economic Development Planning.* Some of the journals she has published in are *Economic Development Quarterly, The Review of Black Political Economy, Growth and Change,* the *Journal of Urban Technology, Economic Development Review, Commentary,* the *Journal of Planning Education and Research,* and the *Journal of Planning Literature.*